THE PRACTICES OF HUMAN GENETICS

Sociology of the Sciences

A YEARBOOK – VOLUME XXI – 1997

The titles published in this series are listed at the end of this volume.

THE PRACTICES OF
HUMAN GENETICS

Edited by

MICHAEL FORTUN
Hampshire College, Amherst, Massachusetts, USA

and

EVERETT MENDELSOHN
Harvard University, Cambridge, Massachusetts, USA

KLUWER ACADEMIC PUBLISHERS
DORDRECHT / BOSTON / LONDON

A C.I.P. Catalogue record for this book is available from the Library of Congress.

ISBN 0-7923-5333-1

Published by Kluwer Academic Publishers,
P.O. Box 17, 3300 AA Dordrecht, The Netherlands.

Sold and distributed in North, Central and South America
by Kluwer Academic Publishers,
101 Philip Drive, Norwell, MA 02061, U.S.A.

In all other countries, sold and distributed
by Kluwer Academic Publishers,
P.O. Box 322, 3300 AH Dordrecht, The Netherlands

Printed on acid-free paper

Printed and bound in Great Britain by MPG Books, Bodmin, Cornwall

TABLE OF CONTENTS

TABLE OF CONTENTS

PREFACE

That concern about human genetics is at the top of many lists of issues requiring intense discussion from scientific, political, social, and ethical points of view is today no surprise. It was in the spirit of attempting to establish the basis for intelligent discussion of the issues involved that a group of us gathered at a meeting of the International Society for the History, Philosophy, and Social Studies of Biology in the Summer of 1995 at Brandeis University and began an exploration of these questions in earlier versions of the papers presented here. Our aim was to cross disciplines and jump national boundaries, to be catholic in the methods and approaches taken, and to bring before readers interested in the emerging issues of human genetics well-reasoned, informative, and provocative papers. The initial conference and elements of the editorial work which have followed were generously supported by the Stifterverband für die Deutsche Wissenschaft. We thank Professor Peter Weingart of Bielefeld University for his assistance in gaining this support.

As Editors, we thank the anonymous readers who commented upon and critiqued many of the papers and in turn made each paper a more valuable contribution. We also thank the authors for their understanding and patience.

Michael Fortun
Everett Mendelsohn
Cambridge, MA
September 1998

INTRODUCTION

In 1986, the annual symposium at the venerable Cold Spring Harbor laboratories was devoted to the "Molecular Biology of *Homo sapiens*." This 51st gathering marked not only the beginning of a new half-century of such events but, as James Watson noted in the prefatory remarks to the printed proceedings, the opening of "a new era of enlightenment." Only once in the first fifty years had humans been the subject of a symposium (in 1964), but Watson was certain that this was a topic to be "returned to over and over during the second 50 symposium years."[1]

As to new terrain scientists turn so too do the analysis of science. Now that human molecular genetics has become so active and potential applications so widespread it has come to occupy a focal position in a complex of social issues. Sociologists, historians, anthropologists, philosophers, ethicists, legal scholars, and many others will be scrambling for years to come to keep up with these developments and their implications, and to create new tools for their analysis. The essays in this volume represent an attempt to come to terms with a field marked as much by rapid changes in sophisticated technoscientific knowledge, instruments and technique, as by profound upheavals in the ways we think about "the ethical," "the biological," "the normal," and indeed, "the human" itself. "What are we?" and "what may we become?" are questions undergoing the most intense negotiations – negotiations taking place in the varied and quite heterogeneous spaces of laboratories, government hearing rooms, corporate board rooms, law courts, clinics, homes, cyberspace, and we hope in the pages of books such as this one.

Over the last decade and a half, the analysis of technoscientific *practices* have been assuming a more prominent role among communities of scholars which take science and technology as their subjects – not displacing such traditional categories as epistemology or ethics, but as a new channel for thinking about those categories. In naming this volume "The Practices of Human Genetics," it was our desire to collect essays which carried out detailed, in-depth studies of the complicated, laborious ways – in both current and historical situations – in which new knowledge, new

---

[1]James D. Watson, "Foreword," *Molecular Biology of Homo sapiens: Cold Spring Harbor Symposia in Quantitative Biology* 51 (1986), p. xv.

Michael Fortun and Everett Mendelsohn (eds.), The Practices of Human Genetics, ix–xiv
©1999 *Kluwer Academic Publishers. Printed in Great Britain.*

technical practices and new social arrangements have been achieved. Human genetics is perhaps particularly messy and ambiguous, and certainly highly charged in the fields often referred to simply as "ethical, legal, and social implications." The questions raised and the responses demanded are among the most difficult, inexact, and contentious; ready-made solutions and easy commitments are subject to immediate dissolution, or at the very least profound, difficult, and fundamental questioning. We believe that it is only through the kind of painstaking analyses of situated practices which these essays demonstrate that we will be enabled to continue to re-orient ourselves to respond to the questions of human genetics.

The hybridization that we see occurring now between fields of genetics – indeed, between organisms themselves – in the "model" systems of such different organisms as mice, *C. elegans, Drosophila*, yeast, and humans, is not without its historical precedents. Doris Zallen delves into the intersections between an "ecological genetics" centered around *Lepidoptera* and early work on human blood groups in the U.K. In her narrative, E. B. Ford is at the growing center of a network which included his students, other colleagues, and institutions like the Nuffield Foundation; it was a network which developed and spread Ford's concepts of polymorphisms and "super-genes" into work on Rh blood groups and into new research and educational institutions like the Nuffield Unit of Medical Genetics in the Liverpool University Department of Medicine.

Peter Keating and Alberto Cambrosio describe the automated laboratories within which molecular genetics is increasingly carried out as a "collective of humans and non-humans, inextricably composed of scientists, machines, and technicians." By focussing on what exactly is being automated, they overlay the questions and concerns of actor-network theory in current sociology of science on themes like "deskilling" taken from old sociology of work. Automation in this context centers around strategies of neither mimicry nor replacement, but of substitution: an addition that results in new contexts of corporate relations and interests, new funding mechanisms, new realms of interdisciplinarity, and shifting boundaries between what counts as science and technology.

Studies of the complexities of practice provide much-needed, constant reminders that human genetics is not a monolithic enterprise, and is neither intent on nor capable of subsuming the world under one point of view. Jean-Paul Gaudilliere details what he calls the "multi-layered complex of practices" which in the 1960s build up a heterogeneous network of humans and non-humans, (in this latter instance mice and their retail markets). Examining the challenge presented by the "milk-influence" model to the genetic models of breast cancer Gaudilliere reveals the complexity, multi-directionality, and disagreements that exist even at the core of the most "pro-genetic" of research enterprises.

The pair of papers by Garland Allen and by Dorothy Nelin and Susan Lindee directly raise the social policy contexts within which the concepts and images of human genetics have been deployed. The "Violence Initiative," the program developed by the U.S. National Institutes of Health in 1992, is the primary focus of Allen's work. The program's aim was to identify the cause behind the perceived increase in violent and anti-social behavior among Americans. The program definition relied heavily on the supposition of a genetic or biological basis for the problems. Allen, using a socio-political analytical frame briefly tells the story, locates the activity in the social context of the times and then challenges the attribution of a genetic basis for complex human behaviors. He criticizes both the rigor of some of the science used in asserting the genetic basis and also the manner in which the press and other public media overstressed the genetic claims. This is a point that emerges very strongly in the study by Nelkin and Lindee. But Allen goes further, relying on his years of historical study of eugenics, and sees a continuity from the past of faulty method and conceptualization. Specifically he notes the poor definition of phenotypes, the tendency to reduce complex behaviors to a single entity – the gene, the uncritical use of heritability estimates, the faulty or casual use of non-human behavioral models and the simple genetic reductions, e.g. "the gene for criminality."

The Nelkin–Lindee contribution picks up these points and asks a series of critical questions focusing on why there is such widespread and easy acceptance in the public literature of the idea of genetic causes for behavioral and social problems – genes for alcoholism, crime, aggression and war. They identify what they label "genetic fatalism," a person is a readout of his/her DNA. They point to the appeal to simplicity that seems behind the willingness to explain complex human issues through recourse to the genes. But they importantly note (and imply the need for further research of the sort that Allen has begun) genetic explanations as compared to societal/environmental ones seem to shift with prevailing social agendas. Genetic explanations, while neutralizing success or failure – "its all in the blood/genes", also relieve a society of responsibility: a criminal or star is born, not made. And if this is at odds with the American ideal of the perfectibility of humans or at least their improbability, so much worse for the ideal. The recent book by Richard

Herrnstein and Charles Murray, *The Bell Curve*, they note is a prime example of this fatalistic assumption. Since basic human attributes like intelligence are genetically bound, any social reforms are likely to have limited consequences and that therefore social policy should be guided by these biological constraints. The book was particularly controversial on the American scene in that it strongly implied that there was a basic racial, genetic background to white/black differences in measured intelligence – IQ. Both authors argued elsewhere that such ameliorative programs for poor black and Hispanic children as Head Start were wasted since they faced a genetically caused barrier. The implications one can draw from studies of this sort are that key elements of the older eugenic program are alive and well. The Galtonian message is replayed in newer form: improving society can be achieved only by improving DNA.

PKU (phenylketonuria) detection and therapy has long been accepted as a model of how to deal successfully with a deleterious genetic abnormality. Widescale screening of the newborn population identified those infants at risk for this metabolic disorder. Therapy, if started early, was seen as achieving high levels of success and confirming the medical genetic paradigm. As Diane Paul demonstrates in her paper the whole story seemed to be unproblematic and acceptable both to those who were advocates of genetic determinism and those opposed to that view. The identification of "a" gene for PKU was a boon to genetic explanation of disease, while at the same time the environmentalist opponents could argue convincingly that the expression of a gene can be drastically altered by changing the environment, i.e. the diet. Therefore they could argue convincingly that biology is not destiny. In fact, as Paul ironically points out the argument was even carried over into the debate over the hereditary basis of intelligence: treatment (amelioration) can change outcomes. In this sense the PKU story showed the flaw in genetic determinism. But Paul goes back into the PKU case itself to identify an important flaw in the story itself. First she notes that the historians and other analysts overlooked key elements in their accounts: the theory was not well understood and it was unclear how long the very restrictive diet must be followed, e.g. into adolescence? into adulthood? She further noted that the results of the therapy were marked by uncertainties as to how thorough the cure actually was; the recipients often exhibited below expected mental abilities, demonstrating some of the problems that afflicted the untreated. Secondly, however, she found almost no discussion in the medical literature (one citation in 3000 articles identified) of the financial costs of the standard dietary therapy for PKU.

It was expensive, $4600/year for adult formula alone, not to mention the fairly high costs of some of the foods required. She asks the obvious social question: if screening is mandatory and abnormality identification obligates treatment (which is non-mandatory) who should bear the costs if they are beyond what a family can afford or insurance will cover? PKU becomes a "moral teacher" of a very special sort; there are significant socio-economic implications in genetic medicine and especially in regimes of mandatory mass screening.

Simone Bateman Novaes further explored aspects of genetic screening in the context of medical practice when she conducted field work with an organization involved in spermbanking and artificial insemination; sitting in on meetings on the review board as they established criteria for the genetic screening of sperm donors gave her a unique position for analysis. This actual practice of genetic decision making exposed a series of ethical dilemmas: who should be recipients of gametes be? what is the physician's responsibility when there is a risk of transmission of genetic diseases (and risk is almost always present)? are there any special problems in the use of gametes which physically mark the offspring even if there is no evident health risk, e.g. cleft palate, which is operable, or polydactyly (extra fingers) which are always visible?

While the specific case of Hans Nachtsheim, the German geneticist who emerged from World War II with a seemingly clean non-Nazi record, is now re-evaluated by Ute Deichmann (he was not nearly as clean or uninvolved as earlier believed), the point that emerges from her study is that the "normal practices" of human genetics in Nazi Germany exceeded what we would accept as humane or ethical. But what also emerges is that beliefs which in retrospect are condemned were widely held at the time and not only in Germany. While the Nazis took to extremes enforced sterilization and euthanasia, the practice of sterilizing the eugenically "unfit" was widely advocated and was law in many other places, including many of the states of the United States. It is the context of these genetic judgements, an extremely prejudicial racial policy and widespread misuse of medical experiment and practice, that seemed to expose the prejudicial character of much non-Nazi eugenic theory and practice. It is certainly the Nazi excesses that forced the virtual halt of eugenic advocacy in the wartime and immediate post-war years in much of the rest of the world. As other papers in this volume indicate, however, a latent eugenism is found within much thinking in the human genetics field and has made the development of many programs for human and medical genetics more sensitive and controversial than might be expected. The commitment of the various programs for mapping and sequencing the human genome (U.S., European, etc.) to allocate a small percentage of all funds expended for scientific research on projects on the social, legal and ethical elements of human genetic research, is understandable only in the context of the historical misuses of genetics and the potential for witting or unwitting repetition of the earlier transgressions.

In the conclusion of her paper Joan Fujimura returns to a central focus of the debates surrounding the mapping and sequencing of the human genome – "the production of the normal and the pathological." The fine examination of the processes of developing the data from which the DNA sequences are gathered and interpreted indicates the uncertainties in the theory and practice of constructing sequences. The problematic nature of the establishment of homologies – a key tool used by genome research communities – is shown to have significant consequences in theory development. Fujimura follows with a discussion of representation and standardization in the establishing of molecular sequences and links standardization and normalization to the languages that are used to identify and stigmatize the genetically pathological or abnormal. She closes her provocative commentary by joining Donna Haraway in claiming that "choice of languages is a choice about...commitments...."

Michael Fortun's own paper is in many ways an experiment: in the focus of the topic under examination, *speed*, and in adapting/creating a distinctive literary style. To one of the editors (Everett Mendelson) it has the effect of almost mimicking the field of focus itself. While Fortun demurs from the supposition that he has attempted to "install speed" as a privilege or special explanatory category in the social studies of science, he has brought a focus to the procedures our images created by the practitioners and celebrants of the Human Genome Project. If "Big Science" serves as a useful identifier to an earlier generation of students and critics of the sciences, "speed" will almost certainly emerge as a useful, if only partial, indicator of a critical quality of varying fields of scientific work. Science in the "fast lane" is not a new phenomenon, certainly not in post-WWII era, but it does call attention to itself even as it often makes a call upon disproportionate resources. The sites, sources, and causes of "speed" science will continue to demand special attention from the social examiners of scientific activity. "Why so fast?" will continue to be a critical if only partially understood/answered question.

MODERN BIOLOGICAL DETERMINISM

The Violence Initiative,

The Human Genome Project, and the New Eugenics

GARLAND E. ALLEN

Washington University, St. Louis, Missouri 63130

Introduction

The cover of the November, 1992, issue of the *Journal of NIH Research* shows a howling rhesus monkey clinging to a tree. However, the story inside is not about rhesus monkeys. Rather, it deals with the "Violence Initiative" of the United States government's Department of Health and Human Services, including the National Institutes of Health.[1] The Violence Initiative is a $400 million program designed to apply the tools of biology – particularly organic psychiatry and behavior genetics – to potential criminals, especially black and Latino youth in America's inner cities. Dr. Frederick Goodwin, in 1992 the head of the Alcohol, Drug Abuse, and Mental Health Administration, described the Violence Initiative as "a public health approach to violence," that focused on screening out and treating preventively "violence-prone individuals."[2] According to Goodwin, various studies within the context of the Violence Initiative aim "to design and evaluate psychosocial, psychological, and medical interventions for at-risk children before they become labeled as delinquent or criminal. This is the basic point of it all ... identifying at-risk kids at a very early age before they have become criminalized."[3] According to Goodwin, this is a public health, or medical, approach to the recurrent problem of violence in our society. Estimated by some to comprise a population of 100,000 or more, such at-risk children come predominantly from what Goodwin calls "high-impact urban areas." Goodwin claims that targeted groups would include those in

---

A part of this paper is based on a position paper published by the International Committee Against Racism (InCAR), first published in the spring of 1993.

1

Michael Fortun and Everett Mendelsohn (eds.), The Practices of Human Genetics, 1–23
©1999 *Kluwer Academic Publishers. Printed in Great Britain.*

the inner city, families in which the parents (or other custodial adults) have a low income and a low educational level, or female-headed households – all synonyms, of course, for poor, urban, African-American (or in some areas, Hispanic-American) populations.

According to Goodwin, the Violence Initiative is the federal government's highest priority for fiscal 1994. One of Goodwin's public remarks indicates the basic approach that the "public health, or medical, model" would take:

If you look at other primates in nature – male primates in nature – you find that even with our violent society we are doing very well. If you look, for example, at male monkeys, especially in the wild, roughly half of them survive to adulthood. The other half die by violence. That is the natural way of it for males, to knock each other off and in fact, there are some interesting evolutionary implications of that because the same hyperaggressive monkeys who kill each other are also hypersexual, so they populate more and reproduce more to offset the fact that half of them are dying.[4]

For his purposes, it is a good thing that Goodwin did not bother to check with any primatologists, because he would have found that there is no substantiation for claiming that in nature half of male rhesus monkeys die violently at the hands of other males.[5] Moreover, there is no evidence that males designated as "hyperactive" necessarily reproduce more offspring (though they may copulate more frequently). Goodwin goes on to offer the "insights" he claims to have about today's urban problems from this bit of pop ethology:

Now, one could say that if some of the loss of social structure in this society and particularly within the high-impact inner city areas, has removed some of the civilizing evolutionary things [influence (?)] that we have built up ... maybe it isn't just a careless use of the word when people call certain areas of certain cities "jungles" – that we may have gone back to what might be more *natural* without all the social controls that we have imposed on ourselves over thousands of years in our evolution.[6]

To complete the circle of logic, Goodwin then suggests that one more part of the Violence Initiative should be an NIH-funded study of rhesus monkey behavior in nature and in the laboratory as a model for how to deal with inner city youth.[7]

In May 1992, when he addressed the American Psychiatric Association, Goodwin presented further details of what research into biological factors in violent behavior would entail.[8] He suggested that elementary schools in "high-impact urban areas" could be used as testing grounds for the first stage of a "triage" system to identify children that might become violent as teenagers or young adults. Elementary school teachers would be asked to identify 12% to 15% of the children who showed characteristics of "early irritability and uncooperativeness." That pool of potential violence-prone children would then be involved in follow-up studies, including psychiatric screening of the family via telephone and then in-person interviews with mental health experts. Apparently, thinking and planning is far enough along on the Violence Initiative to have yielded cost-estimates for the screening program: 7 cents per student for teacher screening and $7 per family for the phone and personal interviews. Once identified, the fate of violence-prone children is not spelled out clearly by Goodwin in either talk, but several suggestions have an ominous ring, such as "day camps" for younger children from poor environments and the possibility of using mood controlling drugs. The latter can be inferred from Goodwin's reference in the May 1992 talk, when he mentioned "serotonergic biochemical imbalances in the brain as useful biochemical markers ... for potential violence"[9]. Given the current, widespread use of amphetamines and methylphenidate (Ritalin) being given to more than 1 million school children every year to control "hyperactivity," as well as newer drugs such as fluoxetine (Prozac), one possible outcome of the Violence Initiative could be the wholesale drugging of inner city youth. That the pharmaceutical industry casts an approving eye on such proposals should be no surprise, since Ritalin, amphetamines, and Prozac are among their most profitable and biggest-selling products.

Predictably, when news of the Violence Initiative and Goodwin's comments became public in mid-May 1992, they produced a vigorous response. Dr. Peter Breggin, a Bethesda-based psychiatrist, author of *Toxic Psychiatry* and a long-standing opponent of biological psychiatry (he heads his own watch-dog operation, the Center for the Study of Psychiatry), attacked the program publicly at a press conference.[10] Newspapers carrying both Goodwin's original statements and Breggin's criticisms produced a flurry of journalistic activity in the summer of 1992.[11] As a result of the controversy, Goodwin was relieved of his position at the helm head of the Alcohol, Drug Abuse, and Mental Health Administration and was "demoted" to head of the National Institute of Mental Health (NIMH). Countering the criticism that he was simply advocating old-fashioned biological determinism, Goodwin claimed that the Violence Initiative is not just raw genetics but a multidisciplinary approach to youth violence: partly biological, based on studies showing that there are modest genetic and neurophysiological factors that predispose some individuals towards violence; partly behavioral, using rhesus monkeys as animal models; and partly sociological, drawing heavily on recent research in psychology and criminology.[12] But it is clear from Goodwin's remarks and other documents that biological components are given a special prominence in the research programs that make up the Violence Initiative.

A second stage in the furor surrounding early exposure of the Violence Initiative was the cancelling of a conference originally scheduled for October 1992, on "Genetic Factors in Crime: Findings, Uses, and Implications," sponsored by the University of Maryland's Institute for Philosophy in Public Policy, with a major grant from the NIH. Responding to public and congressional pressure over the possible connection between the conference and the Violence Initiative, in early September, NIH director Bernadine Healy withdrew funds for the meeting.[13] Much hue and cry from both sides attended this cancellation of the conference, which had been planned originally to bring together both the proponents and the critics of biological and specifically genetic theories of criminal and violent behavior. As it turned out the conference was not part of the Violence Initiative per se but had received its funds from the NIH Program on the Ethical, Legal, and Social Impact of the Human Genome Project, or ELSI. Not only the conference but also the entire Violence Initiative has been labeled racist by psychiatrists and health care workers – people like Breggin – as well as by politicians such as Representative John Conyers, Jr. (D, Michigan) and the memberships of the American Psychological Association and the National Association of Social Workers.[14] In response to critics, officials of the Departments of Health and Human Services (HHS), including NIH, claimed that nowhere have any of their proposals mentioned either race or ethnic groups as having a greater potential for biologically determined criminality.[15] Furthermore, with the order from NIH director Bernadine Healy to freeze funds for the University of Maryland conference, researchers outside as well as inside the NIH felt that their academic freedom had been seriously compromised. For many, Healy's decision represented an abdication of professional responsibility on the part of NIH, a response more to political pressure from the community, especially minority groups, than to the peer review process to which the agency is supposedly committed.[16] At the moment, the status of the Violence Initiative seems assured, with a recent report claiming that NIH funding for the Initiative has been *increased* by $58 million, as well as an announcement from the NSF in December 1994 that it would begin accepting proposals for a $12 million violence-research consortium.[17] Given the increased awareness of violence in the United States, it is unlikely that the entire undertaking will be scrapped. (Funds for the Maryland conference on crime were restored and the meeting was held, still drawing much vocal protest, in September, 1996).

In the following pages, I want to examine some aspects of the Violence Initiative. In doing so, I would like to ask two major questions: (1) For what social and political reasons is an emphasis on the biological foundation of crime and violence being raised at the present time? (2) What can the nature of the present research tell us about the role of the natural sciences in modern political discourse – in particular with respect to the Human Genome Project and the shaping of public policy about violence and crime? To help understand some of the factors that might be responsible for the resurgence of genetic-behavioral determinism, I will compare the present case with the development of what I consider to be a similar one – the Eugenics Movement – during the early decades of this century. The parallels that I will try to draw may be instructive in understanding something of the social forces that are bringing these issues to the surface today. Finally, I would like to ask a third question: What are the implications of the Violence Initiative, in conjunction with the Human Genome Project, for U. S. society in the years ahead?

The Violence Initiative: Science and Its Social Context

The Violence Initiative is not merely the brainchild of Frederick Goodwin or even of the NIH. It is an increasingly organized effort among a number of U.S. government agencies to coordinate research and to propose programs in response to increasing problems of violence in this country.[18] In November 1992, the National Research Council, acting under a directive from the National Academy of Sciences, issued a 400-page report, *Understanding and Preventing Violence*. The study was funded in part by three other federal agencies: The Centers for Disease Control (CDC) in Atlanta, the Justice Department, and the National Science Foundation (NSF). Breggin and Ross-Breggin report that one panel member who helped write the study said its most significant accomplishment is "the unparalleled opportunity to examine the relationship between biomedical variables and violent behavior."[19] While discussions of social science research make up the bulk of the NRC report, biological research was stressed as providing new and important methods for understanding the roots of violent behavior. In addition, the CDC submitted a proposal in June 1992 to coordinate the efforts of several federal agencies into what they termed the "Youth Violence Prevention Initiative". Again, this proposal emphasizes the new perspectives that biological research could provide in solving the age-old problem of crime.

It is no surprise that the Violence Initiative came into being just after the Rodney King riots in Los Angeles, in March 1991. However, the Violence Initiative should not be seen primarily a response to a single event but to a more general economic and social trend that has been developing for over a decade. For example, the Bush administration in 1992 admitted that poverty had reached its highest rate in 27 years. Wages and benefits in many job areas have been severely cut in the past decade, while true unemployment or partial unemployment has increased steadily. In the same time period, drug-related crimes and associated gang violence have increased dramatically in most large U.S. cities. According to the Bureau of Justice Statistics, the number of prisoners convicted of violent crimes has risen from fewer than 25,000 in 1960 to more than 75,000 in 1990. At the same time, the number convicted for drug-related crimes has climbed from fewer than 5,000 in 1960 to more than 100,000 in 1990. Moreover, convictions for crimes of public disorder have increased from 6,000 in 1976 to 25,000 in 1990. All of these increases have differentially affected poor and, in particular, African-American or Mexican-American populations.[20] The challenge that proponents of the Violence Initiative face, of course, is to explain the causes of these increases in violent and antisocial behavior, with the goal of being able to prevent them in the future. In this context, the line gets redrawn between the old alternatives: *nature* versus *nurture*. Clearly, the Violence Initiative speaks to the nature side of the argument.

Starting in 1985, long before the Violence Initiative was even proposed, a theory for the genetic basis of criminality received considerable attention in a new book, *Crime and Human Nature*, by James Q. Wilson and Richard J. Herrnstein, both well-known professors, of government and psychology, respectively, at Harvard University.[21] The book got a lot of attention from the news media, and both authors were interviewed on radio and TV as well.[22] In addition, within the last five years, a host of other human social behaviors have been claimed to have a genetic basis, including alcoholism, manic depression, schizophrenia, general personality factors, risk taking, homosexuality, and shyness. The general reader has thus been treated to a barrage of articles, many of them carried as cover stories by the nation's leading magazines and newspapers (*Time, U.S. News and World Report, Newsweek, The New Republic,* the *New York Times,* and the *Wall street Journal*).

Rarely are the difficulties in assessing the genetic component of complex behaviors spelled out in full in these popular accounts, and the impression that the vast majority of such presentations leave is that new and solid evidence exists for nature over nurture as the cause of violent human behavior. Indeed, an article on the Violence Initiative in the *New York Times* in November 1992 claimed that the various government proposals coming from Goodwin and others were based on "new findings in genetics, biology, and neurobiology."[23] As might be expected, the so-called "new findings" are neither so new nor so convincing as their promoters would like people to believe. Virtually all of the studies purporting to find a genetic basis for complex human behaviors have been reviewed and summarized by Gregory Carey in a paper commissioned by the National Academy of Sciences and published in the volume *Understanding and Preventing Violence*.[24] Carey's overall conclusion is that existing studies are inconclusive, and that, while a significant genetic component to such behavior cannot be ruled out, it is also true that no clear role for genes in human social behavior has yet been demonstrated.

As an example of the sort of "new" evidence that has been forthcoming about the genetics of crime, consider the following two cases, both published between 1992 and 1993. In January 1992, Dr. Allen Beck, a demographer for the Bureau of Justice Statistics, a branch of the Justice Department, compiled data on incarcerated criminals and their families.[25] Beck's survey, which covered more than 2,621 criminals from across the country for the year 1987, found that more than half of all juvenile delinquents, and more than one-third of adult criminals in local jails and state institutions have immediate family members who have also been incarcerated at some time in the past. According to the *New York Times* summary of this research, some criminologists have claimed that Beck's study provides the hard data necessary to demonstrate that criminality runs in families. What is striking is the ease with which such an obvious finding is used to support a genetic argument. Indeed, Richard Herrnstein has claimed that the data are "startling proof" that a significant component of criminal behavior is genetic. It is obvious that such statistics say *nothing* abut what *causes* the incarceration of significant numbers of related people: the sociological explanation is just as likely to be true as the biological. As Dr. Marvin E. Wolfgang, professor of criminal law at the University of Pennsylvania, points out, "Most of these people [in prisons] come from low socioeconomic backgrounds, disadvantaged neighborhoods, where a high percentage of people will be sent to jail whether they are related or not."[26] At least two letters to the editor in a subsequent issue of the *New York Times* (February 14) make the same point.[27] The very fact that such a study can be put forward, as well as accepted in some circles as serious scientific evidence for a genetic component of behavior, attests to the lack of rigor that runs rampant in both the research community and the popular press when treating such issues. That the media gets away with such claims among the reading public may also attest to the scientific illiteracy that our educational system has engendered.

The second example comes from a June 1993 issue of the *American Journal of Human Genetics*. It begins with an anecdotal account of a Dutch school teacher, who, in the early 1960s, traced to a biological cause what he considered to be a pattern of violent behavior among males in his family.[28] A team of Dutch and American researchers examined some of the so-called violent members of the family and found that these individuals had an abnormal gene for monomine oxidase A, an enzyme that usually breaks down several neurotransmitters, including serotonin and dopamine, in the brain. Accumulation of neurotransmitters is thought to be one of the possible causes of violent behavior. To their credit, the researchers issued some cautions about how to interpret their findings. They pointed out that detection of a single gene does not mean that everyone who possesses the gene will necessarily show the behavior: indeed, one member of the family with the gene has not shown any violent behavior in years, while his brother, who also has the gene, has repeated aggressive outbursts. The researchers also admitted that their study does not provide any way of determining the cause-effect relationship between the defective gene and aggressive or violent behavior. Accumulation of neurotransmitters may be involved, but the study provides no evidence whatsoever as to whether increased levels of neurotransmitter are the result of defective genes or of changes in the individual's external environment (neurotransmitter production is highly dependent on levels of external stimulation).

Moreover, in all cases documented, the so-called violent or aggressive behavior was what neurobiologist Steven Rose of the Open University, in England, has called an "arbitrary agglomeration of behaviors, lumping together many reified interactions as all exemplars of the same thing."[29] The various "violent" activities included impulsive aggression, arson, attempted rape, and exhibitionism, activities carried out in different parts of the country at different times and under different circumstances. The lumping together as "aggressive behavior" of such a wide variety of activities – all of which are described from anecdotes – is both arbitrary and misleading. Indeed, additional information revealed that at least several individuals showed their unusual behavior to be in response to a variety of external provocations, such as a negative job evaluation or the death of a relative.

Jonathan Beckwith of Harvard Medical School has also severely criticized the Dutch study. Although the original article is filled with large numbers of measurements of monomine oxidase A and B levels and estimates of genetic marker locations, Beckwith notes that there is practically no detailed information on family history, nor on the context of, or the nature of the violent outburst exhibited by afflicted individuals. Indeed, as Beckwith puts it, "There were plenty of statistics and numbers about the mapping of the gene, but there was basically no information that one could evaluate about whether the people were truly aggressive."[30] Despite the reservations on the part of both researchers and critics, however, the *Los Angeles Times* announced the work to its lay readers with the headline, "Researchers Link Gene to Aggression."

Indeed, as these two cases, as well as the more extensive survey by Gregory Carey show, the new reports of a significant genetic basis for criminal and other social behavior share the same conceptual and methodological problems that have characterized this type of work in the past. Some of these problems include: (1) poorly defined phenotypes – for example, a criminal act is defined by the rules and mores of a given society at a given point in history so that the same behavior can be considered criminal in one circumstance and acceptable or even laudable in another; (2) reduction of complex processes, such as aggressive or violent behavior to a single entity – that is, a criminal or violent act is not a single event or process but is made up of a complex of intellectual, emotional, and behavioral components that cannot be reduced to a single behavioral "thing" and thus to a single unit of study as would be expected in standard genetic analysis; (3) the use of twin and adoption studies, which experience shows entail the interaction of many variables, most of which cannot be accurately assessed; (4) reliance on the use of heritability estimates – that is, the statistical analysis of variance within a population – which does not say anything directly about the genetic components involved; (5) the uncritical and selective use of information from other sciences, especially behavioral models with nonhuman animals, such as primates, where superficially similar behaviors are many times viewed as homologs to human behavior; and (6) the tendency in all such research to resort to extreme forms of genetic reductionism, including references to "*the gene* for criminality" or "*the gene* for shyness" (emphasis added), a level of simplification that geneticists know does not apply even to relatively simple morphological traits, much less to highly complex and plastic traits like social behavior.

It is, perhaps, relevant to note that, over the past several decades, beginning with the work of Arthur Jensen on the genetics of racial differences in IQ in the late 1960s, virtually none of those investigators propounding theories of a genetic basis for human social behaviors have actually been trained geneticists. Their naïveté about making genetical analyses and corresponding claims of genetic causality is thus all the more blatant because they would not stand up to any standard genetic scrutiny. Despite this, however, the overwhelming impression gained within certain segments of the scientific, especially psychological and psychiatric, communities, as well as in the popular media, is that new and more accurate data have been obtained indicating a genetic basis for violent, criminal, and other social behavior.

In addition to the specific claims that certain genes determine at least a strong predisposition to criminality, the whole enterprise of attributing a variety of social behaviors to genetic causes has resulted from the considerable public hype given to genetics in recent years with the advent of the Human Genome Project. Referred to repeatedly as biology's "Manhattan Project," the HGP is unquestionably the largest, organized, government-funded research project in the history of the biomedical sciences. In arguing for such large-scale public support before Congress and in the press, proponents of the HGP, such as James D. Watson, director of the Laboratory of Genetics at Cold Spring Harbor, New York, and Daniel Koshland, editor of *Science*, have claimed that the project will solve every conceivable sort of medical problem, from specific genetic diseases like diabetes, phenylketonuria, or PKU, to major social and behavioral problems like criminality and homelessness. As Watson has said, "We used to think our fate was in the stars. Now we know it is in our genes."[31] Even more blatantly, in the October 12, 1990 issue of *Science*, Koshland, a biochemist by training, wrote:

It is time the world recognizes that the brain is an organ like any other organ... and that it can go wrong not only as a result of abuse, but also because of hereditary defects utterly unrelated to environmental influences... The irrational output of a faulty brain is like the faulty wiring of a computer in which failure is caused not by the information fed into the computer, but by incorrect processing of that information after it enters the black box.[32]

Given the many genuine findings that have already emerged from various areas of molecular genetics, as well as the prospect that at least some additional evidence of importance will result from the work of the HGP, the attempt to generate widespread belief in a genetic basis of social behaviors by linking them to the HGP is particularly important. At the level of popular public discourse, there is a synergistic interaction taking place between molecular genetics and genetic theories of human behavior that gives the latter the appearance of increased credibility.

In summary, then, the context in which the current barrage of theories about genetic basis of social behavior are occurring includes: (1) a persistence of economic and social problems such as unemployment, wage and benefit cuts, and increased urban violence, along with (2) a barrage of publicity about the great medical and psychiatric benefits to come from advances in molecular genetics, in particular those associated with the HGP.

Now I want to turn to an analysis of an earlier case study that has many similarities to the present: the rise of the "science" of eugenics in the first three decades of the present century. This case study can be useful in prompting questions about the social context in which genetic studies of human behavior arise, as well as the political consequences to which they can lead, especially when they are widely accepted in the political arena.

Development of Eugenics in America, 1900–1940

Since much has been written about the history of eugenics in recent years,[33] I will present only a brief overview of the movement with respect to the United States, in order to determine in what ways its history may be instructive for understanding the development of the theories of genetic determinism, and the Violence Initiative today.

The term "eugenics" was coined in 1883 by Francis Galton, first cousin of Charles Darwin and himself a pioneer in the application of statistical and quantitative measurements to biological problems.[34] For Galton, eugenics meant "purely born," referring to his aim of improving the overall hereditary quality of the human species by planned breeding. Although Galton never developed an active eugenics program of his own, his disciples, including Karl Pearson in Great Britain and Charles B. Davenport in the United States, pursued active research programs in eugenics. In the United States, Davenport established The Station for the Experimental Study of Evolution at Cold Spring Harbor, Long Island in 1904, and the Eugenics Record Office in 1910. The former was funded by the Carnegie Institute of Washington, and the latter, initially by Mrs. E. H. Harriman, widow of the railroad magnate. After 1917, the Carnegie Institute of Washington funded both.[35]

Building on Mendel's newly rediscovered principles of heredity in 1900, Davenport and many eugenecists claimed that social traits such as pauperism, manic depression, scholastic ability, feeblemindedness, degeneracy, epilepsy, and criminality were determined by one or two pairs of Mendelian "factors," or genes. Using primarily family pedigree studies, eugenicists in the United States traced through numerous generations what appeared to be hereditary patterns for many forms of "social degeneracy". Training a large number of eugenic field workers through summer courses at the Eugenics Record Office, Davenport and his superintendent, Harry H. Laughlin, collected information on thousands of families, beginning usually with members who were in state mental or penal institutions. Eugenicists used these data, along with self-administered questionnaires sent to college students and other select groups, to argue for a strong hereditarian basis for most social traits.

Among the many social traits thought to have a hereditary base was criminality. Significant increases in the crime rate in the larger cities of the United States at the time, particularly New York and Chicago, had raised concern among social workers, police, and political figures, as well as among the socially concerned middle and upper classes. For example, between 1850 and 1890, the population of the country as a whole had increased 170%, while the criminal population had increased by 445%. In Pennsylvania between 1880 and 1890, the number of criminals incarcerated in the state's penal institutions doubled, increasing the cost of operating such institutions by an alarming 35%.[36] In addition, there was a social cost of crime in terms of lost property, the tracing and apprehension of criminals, the cost of court trials and of incarceration, etc. In one paper at the annual meeting of the National Prison Association in 1900, a distinguished New York lawyer, Eugene Smith, estimated that the annual expense to the nation's taxpayers from various forms of criminal activity was $600 million.[37] By 1910, the Massachusetts Prison Association estimated that the cost of dealing with crime was the second largest item in the state budget, behind public education.[38] Crime was not only a social and a moral problem; it was also an economic one that could no longer be overlooked.

Davenport himself was particularly interested in the genetic basis of criminality, a subject on which he wrote several papers.[39] Laughlin, under the auspices of the Eugenics Record Office, worked closely with Judge Harry Olson, chief justice of the Municipal Court of Chicago, on the effects of alien crime in the United States. Olson, a committed eugenicist, had founded the Psychopathic Laboratory in Chicago to apply scientific methods to the field of criminology. In 1935, Laughlin and Olson published a report of their findings: data on the rate at which crime had increased in the previous decades, the race and nationality of offenders, and the biological, specifically genetic, basis for crime.[40] Lamenting that many political discussions of crime (such as those recently organized by the governors of New Jersey and New York in 1935) virtually ignored the biological basis of criminal behavior, Olson and Laughlin claim that:

Crime is a social defect based on mental defect, and the mental defect is typical of two great divisions of the mind – the intelligence and the emotions. Instead of being sporadic, appearing by chance, the accumulative records show that, like all physical and mental traits, it runs in family stocks and is subject to the laws of genetics like other characteristics.[41]

For Laughlin and Olson, the increase in crime so apparent in the late 1920s and early 1930s was viewed as an index of poor heredity. They state: "Behind the criminal is a problem child; behind the problem child is the inadequate home and the problem parent; and behind the problem parent is poor heredity, which indicates that the whole sequence lies usually in an environment of economic insecurity."[42] Laughlin and Olson did not discount environment altogether, but it is clearly put in a secondary place: "The causes producing crime work from within (biological) and from without (social). Hereditary impulses are primary; the influences of environment are secondary and operate chiefly through temperamental instability which makes it difficult for the individual to withstand the pressure of his environment."[43] For Laughlin and Olson, many social and legal experts mix up cause and effect by claiming that poverty causes crime. "Do the slums make the man or does the man make the slums", they ask. Referring to the Brock Report, a recent report on the rise of crime, Laughlin and Olson argue that "a considerable proportion of defectives do come from slum surroundings but inquiries into the family history of such cases show that in the majority there is evidence of morbid inheritance.... . The inefficiency of the defective tends to depress him to the lowest economic level."[44] Citing anthropologist E. A. Hooton of Harvard, Olson and Laughlin argue, "it can be stated positively that the biological inferiority of the criminal is no less marked than his economic ineffectiveness and his general stupidity. We need a biological New Deal which will segregate and sterilize the anti-social and mentally unfit."[45]

Among criminals and the violently prone, Laughlin and Olson presented a breakdown by racial and ethnic/national origins. A 1926 study of 2,000 delinquents in Chicago, for example, showed that 75% had foreign born parents, with Italians and Slavs having the highest representation. By 1932, a similar survey showed that African Americans ("native Negroes," as Laughlin and Olson called them) Mexicans, and Puerto Ricans had all been added to the list as far overrepresented among the prison population compared with the population at large.[46] The lesson that Laughlin and Olson drew from this message was clear: all the non-Anglo Saxon groups are to varying degrees defective and supply a large percentage of society's criminals. Quoting Charles W. Burr, M.D., Professor Emeritus of Mental Diseases at the University of Pennsylvania, they summarized their findings in as clearly hereditarian a vein as possible: "Crime is not largely the product of economic stress. The criminal, as a psychiatrist defines the word, is born and not made....they are not the product of the slums. They are incurable because they are not suffering from external stress and strain but from an inherent defect in protoplasm."[47]

Even before publication of these results, Laughlin and Olson had been convinced that there was a significant alien component to the increase in crime experienced in most major U.S. cities, especially in the period after

World War I. Through Davenport, the Eugenics Record Office had established close contacts with the Immigration Restriction League, founded in
Boston in the 1890s by Robert De Courcy Ward and Leverett Saltonstall.[48]
The changing pattern of immigration from the 1880s onward had been
particularly disturbing to many white, Anglo-Saxon Protestants, since it had
produced a dramatic increase in immigrants from Southern Europe (Italy in
particular), central and eastern Europe (Poland), the Mediterranean countries
(Turkey, the Balkans, and Greece) and Russia (both before, but especially
after, the Bolshevik Revolution of 1917). Since many of the "new" immigrants were Jewish, the call for immigration restriction took on an increasingly anti-Semitic overtone by the time of World War I. After the war, the
influx of immigrants increased dramatically, due in large part to economic
chaos in many European countries, as well as to the United States' open
door policy. Combined with the return of U.S. soldiers, post World War I
immigration had especially dramatic economic consequences: large-scale
unemployment, increased union membership and, most important, increased
numbers and size of union-led strikes. Foreigners were blamed for much of
the agitation, while their leadership roles in many trade unions targeted
them as a particular menace.[49]

To deal with the increasing problem of immigration between 1920 and
1923, the House Committee on Immigration and Naturalization, under
Albert Johnson, a rabidly anti-union, anti-communist and restrictionist
congressman from Washington, held a series of hearings on immigration
restriction. Johnson and other nativists did not want to restrict immigration
across the board but wanted to restrict selectively the "new" immigrants
from southern and eastern Europe and the Mediterranean countries. Johnson
called on Harry Laughlin as honorary president of the Eugenics Research
Association, to become "expert eugenic witness" to the House Committee
and to testify before the committee about the relative genetic merit of
different racial and ethnic groups. Laughlin's testimony covered a period of
several days during two different committee meetings, on April 16 and 17,
1920.[50] With charts and a variety of statistics, he drove home the point that
the so-called "new immigrants were genetically inferior to the older, Anglo-
Saxon, and Nordic immigrants of the nineteenth century and that they were
the source of many of the nation's most glaring social and economic
problems.

Prominent among the genetically determined social traits were criminality and violent behavior. Although his studies with Judge Olson on
immigrant crime in Chicago were not completed in time for the final series
of hearings on immigration restriction in 1924, Laughlin demonstrated that
in virtually all state penal institutions he had surveyed, the foreign-born

were dramatically overrepresented compared with their numbers in the population at large. Partly as a result of Laughlin's "scientific" testimony, restrictionist sentiment carried the day, and Congress passed the Johnson Act (Immigration Restriction Act) in 1924. The Johnson Act selectively limited immigration from southern and eastern Europe, the Mediterranean countries, the Balkans, Russia, and, in general (irrespective of geography), people of Jewish descent.[51]

In addition to immigration restriction, eugenicists such as Laughlin lobbied strongly for eugenical sterilization laws that would include, among other "defectives," repeat offenders in state penal institutions. With Davenport's support, Laughlin drew up a model sterilization law that could be introduced, with modifications to meet local needs, into the various state legislatures. Although the first law for eugenical sterilization (as opposed to strictly punitive measures) had been introduced in Indiana as early as 1907, a widespread movement to legalize involuntary sterilization did not develop serious momentum until World War I.[52] Throughout the 1920s and 1930s, Laughlin and his associates campaigned hard in a large number of state legislatures, arguing that sterilization was an efficient and "progressive" way to stop the spread of crime – at its roots.[53] By 1935, Laughlin could boast that more than 30 states had adopted such legislation and that another 10 were considering it seriously.[54] And, by 1941, more than 33,000 people (60% of them women) had been sterilized on the basis of these laws (the number had exceeded 60,000 by the late 1970s, when most of the laws had been repealed).[55]

When the constitutionality of eugenical sterilization laws was tested in Virginia in 1925, in the now-famous Buck *vs* Bell case, Laughlin served as one of the two primary expert eugenics witnesses to the court (the other was E.A. Estabrook, one of Laughlin's and Davenport's colleagues at Cold Spring Harbor).[56] Without ever meeting or interviewing the plaintiff, Carrie M. Buck, Laughlin certified that she was a "low-grade moron" with a "mental age of nine" and "the potential parent of socially inadequate offspring." A few years later, Laughlin's influence extended even further, when his model eugenical sterilization law was used as the basis for the Nazi sterilization laws of 1933.[57]

Although it is clear that the U.S. sterilization, and especially the immigration, laws would probably have passed whether or not eugenicists had been involved, the claim for a biological basis for undesirable social traits made the legislative route far smoother. In Germany, as a number of recent historians of science have shown, biological arguments helped prepare the way for the atrocities of the Holocaust.[58] Not wishing to distort history nor to indulge in hyperbole, I do want to stress, however, the importance of recognizing the role that biological and in particular genetic arguments, even when they appear in rather innocuous technical garb, can play in the much larger social and political arenas. In most of the formulations in which it appeared in the past as well as in the present, the phrase "Biology is destiny!" has led to disastrous human consequences.

Conclusion

There are many points of similarity between the social context of the eugenics movement in the early decades of the twentieth century and the wave of genetic determinism circulating in recent years, in the 1980s and 1990s. For one thing, both share the context of an economic downswing that in their respective periods seems to show no indication of improvement. For example, unemployment hit major peaks in 1914, 1921, and 1932. The early decades of the twentieth century saw much violence, especially that associated with the "labor wars," that is, the confrontation between capital and labor with regard to union organizing and collective bargaining. Numerous strikes occurred on a massive scale from the 1870s onward, signs of the much larger, ongoing labor struggle.[59] No wonder Laughlin, Davenport, and other eugenicists in the United States were concerned to show that those who committed violent crimes were genetically defective. Indeed, Laughlin and Judge Olson saw the "lawlessness" that was sweeping the nation to be a product of "our historical past (wild west), our racial mixture, and our industrial conflicts."[60]

While the external character of the economic and social contexts of the 1980s and 1990s differs from that of earlier years, the core problems have much in common. Both periods involve a sense of economic unpredictability, increase in under- or unemployment and cutbacks in pay, a sense of social dislocation and deterioration. Both periods witnessed an increase in the number of people in state-dependent institutions such as mental hospitals or prisons, an increase in concern about immigrants taking jobs from native residents, and attacks on welfare programs and the "dependent classes" (see, for example, the article on immigration in the *Wall Street Journal* of July 7, 1993). In charting the course of state sterilization laws, for example, Philip Reilly suggests that one major factor in the remarkable increase in the number of states adopting such laws in the period 1927–1935, was the severe budget cuts that they faced during the Depression.[61] Similarly, Paul Lombardo credits a major recession in the state of Virginia in 1923–1924 with helping overcome opposition in the legislature to the passage of a eugenical sterilization law.[62] A similarly interesting parallel is the fact that in both cases the claims for genetic determinism followed rapid and path-breaking developments in the laboratory genetics of the day: the old eugenics movement developed in the wake of the rediscovery of and expansion of Mendel's concepts of heredity, while the new movement is taking advantage of rapid developments in molecular genetics, particularly those associated with the Human Genome Project.

Although their similarities are many, the new and old arguments for genetic determinism are also different in some ways. In the past, eugenicists were concerned primarily with legislating and controlling breeding patterns, while modern advocates focus more on a medical model for dealing with deviant behavior, including treatment by drugs, and should the technology become available, gene therapy. On a more technical level, today's genetic determinists base their arguments more on statistical analyses of variance between identical twins or siblings in adoption studies, while the older eugenicists based their analyses largely on family pedigrees. However, even today, the "family history" plays an important role in many studies of behavior genetics.

Despite the differences, the question still remains what function, if any, do these two movements share in terms of the political, economic, and social context in which they are embedded? I argue that in times of economic or social crisis, theories of genetic determinism do, in fact, serve very similar functions. They purport to treat complex and otherwise tradi-tionally intractable social problems by what appear to be new "scientific" methods. Especially when the scientific approach is presented in new and highly technical terms, it is much simpler to think of a complex problem, such as increase in crime rate or mental illness, as a result of innate biological causes that do not in any way challenge the economic or social *status quo*. The victim is, under these conditions, the cause of his or her own problems; clearly, the simplest way to deal with such problems is to "fix" them by scientific means, thus preventing their recurrence. Although tracing social problems to genetic causes appears to take the onus off the individual, the ultimate result has always been to marginalize some people and desig-nate them as hopeless incurables. Consequently, any argument claiming that violent or criminal behavior is due to genetics does not guarantee a more humane treatment of stigmatized individuals.

More important, genetic determinist arguments provide the rationale for an economically more "efficient" (that is, less expensive) way of dealing with so-called defective individuals. As costly as they are, today's behavior-control drugs are cheaper than long-term social or psychological therapy, job counseling and job retraining programs (especially if there are no jobs to go to), or programs for reducing stress in the workplace, the schools or at home. The logical outcome of medicalizing behavioral or personality problems is the social control of individuals and the population at large by various biological – physiological, neurobiological, or genetic – manipulations. At a time when people are under economic and social stress, genetic determinists' theories divert attention from the real causes of scarcity or cutbacks and we end up scapegoating one another – by race, ethnic origin, gender, or sexual orientation. Instead of attacking more real and accessible causes of our social problems – the unequal distribution of wealth, for example, and the control of economic resources by a tiny percentage of the population – we blame the lower socioeconomic groups for what is perceived as their own biological deficiency.

Genetic determinists often complain that their critics are motivated by political biases and simply do not want to face the very real possibility that people are not created equal biologically, and that our social and political system has to take these inequalities into account. At the same time, genetic determinists also portray themselves as politically unbiased scientists, following wherever the data may lead, even if the conclusions are not fashionable. I do not deny that my skepticism about such theories is politically motivated, but the politics is not simple, idealized egalitarianism. Rather, my skepticism about and opposition to such theories arises from these separate but interrelated issues: (1) the persistently inconclusive nature of the data; (2) the historical ways in which the theories have been used; and (3) the political biases that *are* embedded in genetic determinist arguments, despite claims to the contrary. To conclude, I would like to examine these three claims more closely.

(1)   The enormous difficulty involved in obtaining data that in any rigorous way can separate the effects of heredity from those of environment in understanding the causes of human behavior ought to make anyone immediately suspicious of strong claims about "the gene for…" or "a significant genetic component to…" any given behavior. Whether measuring cranial index, IQ scores, or serotonin levels, over the years, the determinists have failed to produce data that withstands close scrutiny. Being wary or skeptical about such data does not mean rejecting them out of hand; but given the methodological problems involved and the unbroken string of misfired theories, skepticism rather than naïve acceptance or neutrality seems to be a logical response. A useful and succinct summary of many of the methodological problems associated with current behavior genetics can be found in Steven Rose's "The Rise of Neurogenetic Determinism,"

published in *Nature* in 1995.[63]

(2)　Historically, theories of genetic (biological) determinism have always been used to deny people – almost always the less fortunate and down-trodden of society – access to resources, to limit rather than to expand their options. There is no logic to this outcome. If a trait is indeed determined to be genetic, the social policy deriving from this informa-tion need not lead inevitably to restriction. Down's Syndrome is known to be a genetic condition, but that knowledge has led to two different social policies over the years. Before the 1960s, Down's children were largely institutionalized and accepted as having a limited and rather uniform range of capabilities. Largely for social reasons, policy shifted in the late 1950s, and in the last 30 years, most Down's patients have been raised at home and educated through special schools. Genetic limitations, even when they clearly exist, do not necessarily dictate restrictive social policy. Our social policy derives from our social values, which do not in any way hinge on biological data. The fact that genetic determinist theories have almost always been used to limit who gains access to resources suggests an underlying political bias among the social and political activists, if not among the researchers them-selves, who have advocated and continue to advocate these views.

(3)　It seems naïve, if not ingenuous, for advocates of genetic determinism to claim to be "disinterested scientists," while charging their critics with political bias. We all have biases of various sorts, and they can never be eliminated totally, nor are they always a bad thing. The problem comes when people pretend *not* to have biases, indeed, to be "objective" in some abstract sense. Bias can help provide insights, as well as blind us to alternative views, so it seems to me that the best policy is to bring these biases out in the open and examine them, to find out if they are helping or hindering the work at hand. It is now so commonplace in the history and sociology of science to recognize, and even seek out and analyze, social and political biases in scientific work – the so-called social constructionist views – that it is difficult to believe many practicing scientists still cling to old positivist myths about the neutrality and objectivity of science. But I suppose I would be naïve to assume that historians of science have necessarily made such a large impact on the scientific community. At any rate, propo-nents of genetic determinism have seldom seemed to display less bias in their views than their critics. I believe that the problem is not so much whether or not they have a bias, but that they do not admit what their bias is. There is little to be gained by debating details of statistics or experimental design when large issues of political and social philosophy remain unexamined.

Following on my arguments in this paper, I want to conclude by suggesting that the Violence Initiative is a highly simplistic approach to the complex problems of the increasing violence, frustration, and anger in contemporary U.S. society. Perhaps some people will view my analysis as too simplistic, or no more realistic than the historical theories that I criticize so strongly. But what we see as the solutions available to us depends very much on what we believe the causes to be. If my analysis is correct, that the Violence Initiative is serving much the same function in today's society as eugenics did in its day, then we must monitor the program development and expose its simplistic and unsupported claims wherever they come up.

Notes and References

1. Nancy Touchette, "Cowering Inferno: Clearing the Smoke on Violence Research," *Journal of NIH Research 4* (November 1992), pp. 31–33.
2. Frederick K. Goodwin, "Conduct Disorder as a Precursor to Adult Violence and Substance Abuse: Can the Progression be Halted?" address to the American Psychiatric Association, Washington, D.C., May 5, 1992. Recorded by Mobile Tape Co., Inc., 25061 West Avenue, Stanford, Suite 70, Valencia, CA 91355; see also report of Goodwin's remarks in the *Washington Post*, July 29, 1992, Metro Section, B1.
3. *Ibid.*
4. Goodwin, in speech to National Mental Health Advisory Council, February 11, 1992; see "Partial Transcript of a Draft to the National Mental Health Advisory council," unpublished transcript.
5. For examples of current work underscoring the complexity of primate behaviors (including the rhesus monkey, *Macaca mulatta*), see Barbara B. Smuts, Dorothy L. Cheney, Robert M. Seyfarth, Richard W. Wasserman, and Thomas T. Struhsaker, ed., *Primate Societies* (Chicago, University of Chicago Press, 1987); especially useful is a chapter on the problems of comparing primate and human behavior: Robert A. Hinde, "Can Nonhuman Primates Help Us Understand Human Behavior?" Chapter 33, p. 421.
6. Goodwin, in speech to National Mental Health Advisory Council.
7. *Ibid.*
8. Goodwin, "Conduct Disorders...".
9. *Ibid.*
10. Peter R. Breggin, "The Violence Initiative" – a Racist Biomedical Program for Social Control," *The Rights Tenet* (Summer 1992) p. 3; see also Peter R. Breggin and Ginger Ross-Breggin, "A Biomedical Programme for Urban Violence Control in the U.S.: The dangers of Psychiatric Social Control, *Changes* (March 1993), 59–71, especially p. 63; Peter R. Breggin. *Toxic Psychiatry* (New York., St. Martin's Press, 1991).
11. See, for example, "A Cure for Violence?" *Los Angeles Times*, April 24, 1992, p. E–1,

4; "Hunting," *San Francisco Weekly*, July 15, 1992, pp. 2–3; "Science and Sensitivity: Primates, Politics and the Sudden Debate over the Origins of Human Violence," *Washington Post*, March 1, 1992, p. C–3; "New Storm Brews on Whether Crime Has Roots in Genes," *New York Times*, September 15, 1992, p. C–1; "U.S. Hasn't Given up Linking Genes to Crime" (Letters to the Editor), *New York Times*, September 18, 1992, p. A–34; "Study to Quell Violence Is Racist, Critics Charge," *Detroit Free Press*, November 2, 1992, pp. 1, 11A; "Study Cites Biology's Role in Violent Behavior," *New York Times*, November 13, 1992, p. A–12.

12. Goodwin, interviewed on WAMU Radio, Washington D.C., July 22, 1992.

13. Touchette, *"Cowering Inferno..."* p. 32.

14. *New York Times*, March 8, 1992, p. 34.

15. Touchette, *"Cowering Inferno..."* p. 31.

16. Christopher Anderson, "NIH, Under Fire, Freezes Grant for Conference on Genetics and Crime," *Nature 358*, July 30, 1992, p. 357; see also David Wasserman's response to the cancellation of the conference, "In Defense of a Conference on Genetics and Crime: Assessing the Social Impact of Public Debate," *Chronicle of Higher Education,* September 23, 1991, p. A–44.

17. W. Wayt Gibbs, "Seeking the Criminal Element," *Scientific American, 272*, March, 1995, pp. 100–109, especially p. 106.

18. Breggin and Ross-Breggin, "A Biomedical Programme...," pp. 59–62.

19. *Ibid.*, p. 62.

20. Statistics from *Newsweek* (June 14, 1993): p. 32.

21. James Q. Wilson and Richard Herrnstein, *Crime and Human Nature* (New York, Simon & Schuster, 1985).

22. See, for example, Richard J. Herrnstein and James Q. Wilson, "Are Criminals Made or Born," *New York Times Magazine*, August 4, 1985, *31*; "Genetic Traits Predispose Some to Criminality: A Conversation with James Q. Wilson," *"U.S. News and World Report,* February 10, 1986, p. 67; Also, Wilson was interviewed extensively on National Public Radio (NPR) and other information stations during the months immediately following publication of *Crime and Human Nature*, despite the fact that it was reviewed adversely by prominent psychologists and sociologists including: Leon J. Kamin, *Scientific American, 254* February 1986, pp. 22–27; and J.P. Scott, *Social Biology, 34*, (Nos. 3–4, 198), pp. 256–265. The popular press continued to present the findings as "new" and important evidence for a significant genetic factor in crime.

23. Fox Butterfield, "Study Cites Biology's Role in Violent Behavior," *New York Times*, November 13, 1992, p. A–12.

24. Gregory Carey, "Genetics and Violence," eds. Albert J. Reiss, Laus A. Milzek, and Jeffrey Roth, *Understanding and Preventing Violence, volume 2* (Washington, D.C.: National Academy Press, 1994, pp. 21–58.

25. The recent studies are still unpublished but are summarized in the *New York Times*, January 31, 1992, p. A–1; Earlier studies by Beck have been published as reports of the Bureau of Justice Statistics.

26. *New York Times*, January 31, 1992, p. A–1.

27. Letters to the Editor, *New York Times*, February 14, 1992.

28. H.G. Brunner, M.R. Nelen, P. van Zandvoort, N.G.G.M. Abeling, A.H. van Gennip, E.C. Wolters, M.A. Kuiper, H.H. Ropers, and B.A. van Dost, "X-Linked Borderline Mental Retardation with Prominet Behavioral Disturbance: Phenotype, Genetic

Localization, and Evidence for Disturbed Monamine Metabolism," *American Journal of Human Genetics, 52* (1993), pp. 1032–1039; see also, H.G. Brunner, M. Nelen, X.O. Breakefield, H.H. Ropers, B.A. van Dost, "Abnormal Behavior Associated with a Point Mutation in the Structural Gene for Monamine Oxidase A," *Science, 262,* OXr 22, 1993, pp. 578–580.

29. Steven Rose, "The Rise of Neurogenetic Determination," *Nature, 373,* February 2, 1995, p. 380.
30. Sheryl Strolberg, "Researchers Link Gene to Aggression," *Los Angeles Times,* October 22, 1992, p. A–36.
31. See John Horgan, "Genes and Crime," *Scientific American, 268,* February 1993, p. 26.
32. Daniel E. Koshland, "The Rational Approach to the Irrational" (Editorial), *Science, 250,* October 12, 1990, p. 189.
33. Diane B. Paul, *Controlling Heredity.* Daniel J. Kevles. *In the Name of Eugenics* (New York: Random House, 1985); M.B. Adams ed., *The Wellborn Science* (New York: Oxford University Press, 1992); Garland E. Allen, "Eugenics and American Social History," *Genome, 31* (1989) pp. 885–889; and Kenneth Ludmerer, *Genetics and American Society* (Baltimore: Johns Hopkins University Press, 1972).
34. Francis Galton, *Inquiries into Human Faculty and Its Development* (London: MacMillan & Co., 1883), pp. 24–25.
35. Garland E. Allen. "The Eugenics Record Office at Cold Spring Harbor: An Essay in Institutional History, *Osiris, New Series 5* (1986), pp. 225–264.
36. Estimates by H.M. Boies, *Prisoners and Paupers* (New York: Putnam's, 1893), p. 10, in Philip Reilly, *The Surgical Solution* (Baltimore: Johns Hopkins University Press, 1991), p. 17.
37. Reilly, *The Surgical Solution,* p. 17.
38. *Ibid.*
39. See, for example, Charles B. Davenport, "Crime, Heredity and Environment," *Journal of Heredity, 19,* July 1928,: pp. 307–313.
40. See, for example, the mimeographed pamphlet, "The Biological Basis of Crime" (1929); no author given but containing numerous quotations from Judge Olson and written in the informal, somewhat careless style typical of Laughlin. Numerous copies of the pamphlet were found in the H.H. Laughlin Papers, Pickler Memorial Library, Northeast Missouri State University, Kirksville, Missouri. I am proceeding on the grounds that Laughlin was the primary author but probably collaborated to a considerable extent with Olson.
41. *Ibid.* (Frontispiece).
42. *Ibid.,* p. 5.
43. *Ibid.,* p. 5.
44. *Ibid.,* p. 6.
45. *Ibid.,* p. 7.
46. *Ibid.,* passim.
47. *Ibid.,* p. 9.
48. On the Immigration Restriction League and the growing nativist sentiment in the United States at the end of the nineteenth and beginning of the twentieth centuries, see Barbara Solomon, *Ancestors and Immigrants* (Chicago: University of Chicago Press, 1972); See also Oscar Handlin, *The Uprooted, The Epic Story of the Great Migrations That Made the American People* (Boston: Little, Brown, 2nd ed., 1973); See also John Higham, *Strangers in the Land* (New York: Athenaeum, 1966).
49. Allen Chase, *The Legacy of Malthus* (New York: Basic Books, 1977),: p. 255; see

     also Charles Leinenweber, "The Class and Ethnic Bases of New York City Socialism, 1904–1915," *Labor History, 22* (1982), pp. 31–56.

50. Frances Hassencahl, *"Harry H. Laughlin, Expert Eugenics Agent for the House Committee on Immigration and Naturalization,* 1921 to 1931" Doctoral Dissertation; Case Western Reserve University 1972; Cleveland. See also Kenneth Ludmerer, *Genetics and American Society* (Baltimore: Johns Hopkins University Press, 1972); See also Garland E. Allen, "The Role of Experts in Scientific Controversy," in H.T. Engelhardt and A.L. Caplan, eds., *Scientific Controversies: Case Studies in the Resolution and Closure of Disputes in Science* (Cambridge, England: Cambridge University Press, 1987), p. 183.

51. Allen Chase, *The Legacy of Malthus*, pp. 274–295.

52. Reilly, *The Surgical Solution*, p. x.

53. Harry H. Laughlin. "Further Studies on the Historical and Legal Development of Eugenical Sterilization in the United States," *American association on Mental Deficiency, 41* (1936), pp. 96–110.

54. *Ibid.*, p. 107 (Fig. 2).

55. See summary of sterilization data in Reilly, *The Surgical solution*, p. 97; for the post-1941 figure, see p. 165.

56. Paul A. Lombardo, "Three Generations of Imbeciles Is Enough: New light on Buck vs Bell," *New York University Law Review, 60* (1985), pp. 30–62. Much of Laughlin's and the other "expert" witnesses, testimony is summarized in Harry H. Laughlin, *The Legal Status of Eugenical Sterilization* (Chicago: The Psychopathic Laboratory of the Municipal Court of Chicago, 1929), especially pp. 21–29.

57. Garland E. Allen, "The Eugenics Record Office...," p. 253.

58. Robert Proctor, *Medicine under the Nazis* (Cambridge, Mass.: Harvard University Press, 1988); Benno, Müller-Hill, *Murderous Science* (New York: Oxford University Press, 1988); and Paul Weindling, *Health, Race and German Politics between National Unification and Nazism, 1870–1945* (New York: Cambridge University Press, 1989).

59. Sidney Lens, *The Labor Wars* (New York: Doubleday, 1973).

60. Laughlin and Olson, "The Biological Basis of Crime," p. 5.

61. Reilly, *The Surgical Solution,* pp. 91, 93–94, 101.

62. Lombardo, "Three Generations of Imbeciles...," p. 97.

63. Steven Rose, "The Rise of Neurogenetic Determinism," pp. 380–382.

PROJECTING SPEED GENOMICS

MICHAEL FORTUN

Institute for Science and Interdisciplinary Studies

1. Big Science in the Twilight Zone

The primary vehicle that carries human genetics at the present time is referred to habitually as the "Human Genome Project." Much attention has been given to the question of whether or not the HGP is "Big Science," the form thought to be most characteristic of technoscientific projects in the postwar era. This essay may be thought of as an attempt to map the possible effects of shifting our analytic focus from the spatial to the temporal, to see if "Big Science" might be more productively thought of as "Fast Science."

I have chosen to begin with an image from the immediate postwar decade, a time in the United States characterized by W.T. Lhamon as the invention by poets, novelists, musicians, and Supreme Court justices of a culture of "deliberate speed."[1] A time in which the Central Intelligence Agency was modulating the velocities of the body, searching for new neurochemical means of interrogation that might yield precious insight into an accelerating race between nations freezing into the moving stasis of the Cold War. There, then, "two different drugs with contradictory effects" were administered to create a "twilight zone," where interrogation might proceed with the best chances of success. As Martin Lee and Bruce Shlain describe in their historical study:

CIA doctors attempted to extend the stuporous limbo as long as possible. In order to maintain the delicate balance between consciousness and unconsciousness, an intravenous hook-up was inserted in both the subject's arms. One set of works contained a downer, the other an upper (the classic "goofball" effect); with a mere flick of the finger an interrogator could regulate the flow of chemicals. The idea was to produce a "push" – a sudden [n.b.] outpouring of thoughts, emotions, confidences, and whatnot. Along this line various combinations were tested: Seconal and Dexedrine; Pentothal and Desoxyn; and depending on the whim of the spy in charge, some marijuana (the old OSS standby, which the CIA referred to as "sugar") might be thrown in for good measure.[2]

Michael Fortun and Everett Mendelsohn (eds.), The Practices of Human Genetics, 25–48
©1999 *Kluwer Academic Publishers. Printed in Great Britain.*

The method proved to be too unreliable, resulting more in a stuporous limbo than valuable information. Still, assigning myself the role of a "spy in charge" of interrogating the so-called Human Genome Project, it is a risk I am prepared to take.[3] What new information might we glean by strapping the Human Genome Project into a chair informed by juxtaposed speed vectors, collisions, and caroms of lines of variable speeds, from the excruciatingly slow to the hyperfast?

In this essay, I work primarily with the one IV hookup, the accelerative side, flicking the dial for the slow flow more as a periodic reminder than the equally important constituent of analysis, which it should be. But a first crack at the subject of speed should not bog down in slowness. So in the interest of speeding up the exposition, let me overstate its main story line: the Human Genome Project does not exist. (Having written a history of it, I feel I have some right to say this.[4]) Despite incessant references by journalists, scientists, and many of us in the science studies community to this unitary vehicle signified by those three capital letters HGP, and despite the persistence of such control metaphors as James Watson's or Francis Collin's being "at the helm" or "taking over the reins," the vehicle and its driver are only there in the way that subatomic particles are "there": as a ceaseless flow arrested by the machinic assemblages of the observers for particular rhetorical, that is to say, pragmatic ends. Study of the "Human" has actually dispersed into a zoo blot of research organisms – yeast, mice, *C. elegans, Drosophila* – each of which attracts its own degree of attention and funding, and which hybridize with the "Human" either in the (real) space of the laboratory or in the (virtual) space of the genetic databases. The "Genome," too, loses its distinctly defined boundaries, dispersed into technology itself: what is desired is not so much genomes as it is genomics, the assemblage of techniques, tools, and concepts for producing, analyzing, and manipulating genomes. And the postulated coherence of a localizable, definable "Project," with a manageable, completed endpoint – the "Holy Grail" of a totally mapped, totally sequenced "reference" genome – is a fantasy whose primary function is providing coherence and closure to an epic narrative – a narrative that holds a good deal of exchange value with congressional patrons.

If genomics is seen to be about the circulation of materials, information, skills, capital, products, explanatory powers, and therapies, then what we habitually call the Human Genome Project should be seen only as the acceleration of those patterns of circulation brought into existence well before the mid-1980s. The common journalistic description of the Human Genome Project as "a project to map and sequence the entire human genome" is not only inaccurate, but more important, is incomplete and should at least be revised to read "a project to map and sequence the entire human genome *as fast as we possibly can.*" "Projecting Speed Genomics" would be a more appropriate moniker. In this regard, I share the opinion of people (discussed in more detail later in this paper) who tried to argue that the Human Genome Project was nothing new, just the speeding up of research lines that already existed. The question is, what do the words "only" and "just" mean; Does saying "just speeding up" or "same, only faster," miss new effects – new cycles of circulation, epicycles within old circulation patterns, forces of angular momentum, "spinoffs"?

Given such obvious obsessions with instantaneity as fax machines and other telecommunications technologies, the endless marketing of faster computer chips, stock market fluctuations tied to computerized trading programs, fantasies of high-speed mag-lev trains, thrash and speed metal music, etc., some mapping of Projecting Speed Genomics onto the broader cultural constellation of speed seems both possible and necessary, and would proceed along the lines that Paul Virilio has already begun to lay out:

A capitalism that has become one of jet sets and instant-information banks, actually a whole *social illusion* subordinated to the strategy of the cold war. Let's make no mistake: whether it's drop-outs, the beat generation, automobile drivers, migrant workers, tourists, Olympic champions or travel agents, the military-industrial democracies have made every social category, without distinction, into *unknown soldiers of the order of speeds* – speeds whose hierarchy is controlled more and more each day by the State (headquarters), from the pedestrian to the rocket, from the metabolic to the technological.[5]

While referring to this larger constellation is practically unavoidable, my main goal here is to provide some limited, concrete examples of how speed was talked about during the construction of the Human Genome Project: what it was, how it could be achieved, why it was so incredibly desirable. In taking this more empirical approach, one outcome we can hope for is to arrive at a better sense of the limits of discussions such as Virilio's, discussions that participate too readily in the speed vectors to which they call attention. Such discussions about speed are useful for their sweeping vision of "acceleration as benefiting primarily an invisible, universal, capitalistic power that, through means of telecommunications, produces money – not goods – and is no longer a centralized but a dispersed force."[6] But such discussions call for some supplementation from the nuances, complexity, and contradictions that become more apparent at lower speeds.

Statements about a generalized speed culture raise perhaps the most important question: Is speed a matter of deliberation and choice, or is it some mysteriously irresistible force in itself? This question itself may be framed badly, as was a question directed at me by another analyst of the Human Genome Project when I presented some of this material in a talk: "Well, how fast do *you* think it should go?" he asked. His question assumes a position for the cultural analyst of science as a kind of traffic cop – but what kind of calculus would yield the equivalent of "55 miles per hour" for a socially and ethically responsible genomics? And at what intersection would one place the radar detector, set up the speed trap, stand and raise your hand in an authorized gesture of "Stop!"? No one knows. This has not kept some of the public discourse about the "Human Genome Project" from including such references to an implicit belief in choice and control, however. For example, in 1989, Dr. Robert G. Martin, chief of the microbial genetics section in the Laboratory of Molecular Biology of the National Institute for Diabetes Digestive and Kidney problems, raised many issues in an article critical of the HGP, but speed stood out as one of the biggest issues. "Why the race to complete the mapping of the human genome in 5 years?...I do not deny that mapping the human genome is worthwhile. I do question whether there is justification for the urgency it is receiving...I simply cannot fathom the haste."[7] Or as Harvard microbiologist Bernard Davis asked succinctly in the same issue of *NIHAA Update*: "What's the Big Hurry?"[8] So, while I may agree with the spirit of questioning in these comments – and with someone like Evelyn Fox Keller, when she states that "the real controversy about the Human Genome Project is about pace, it's about the speed. Do we spend $3 billion on it or do we just let it happen a little more slowly?"[9] – nevertheless, it remains unclear to me if (and how) speed in this case can be a matter of public, or even elite, deliberation. I hope that by asking a series of ethnographic questions – How does speed distinguish the Human Genome Project from what came before it, or what might have replaced it? What produces speed, and what limits it? Where are its sources? In what situations is speed "of the essence?" In what situations is is sacrificed for other considerations? – we can better map the space between choice and interpolation, to see where we are, and also, where we might move.

2. Speed Labs

Western man has appeared superior and dominant, despite inferior demographics, because he appeared more rapid. In colonial genocide or ethnocide, he was the *survivor* because he was in fact *super-quick* (*sûr vif*). The French word *vif*, "lively," incorporates at least three meanings: swiftness, speed (*vitesse*), likened to *violence* – sudden force, abrupt edge (*viv force, arête vive*), etc. – and to *life* (*vie*) itself: to be quick means to stay alive (*être vif, c'est être en vie*)![10]

Much of the history of Projecting Speed Genomics in the 1980s appears in the technological development of a genomic armamentarium that has grown to include the polymerase chain reaction (PCR), fluorescent *in situ* hybridization (FISH), souped-up vector vehicles such as cosmids and yeast artificial chromosomes (YACs), pulsed-field gel electrophoresis (PFGE), a dizzying array of highly purified and highly specialized restriction enzymes, and dozens of other cyborg assemblages, including the centerpiece of speed obsession and production, the DNA sequencer.[11] Such standardized tools and biomaterials for genomics research are often advertised on the basis of their ability to create speed, to reduce the time quotient in the equations of efficiency. A new market niche of "downright boring" products and processes has been created as a service sector to the genomics project, in which companies like Bios Laboratories (which claims to be the "top supplier of experimental biologicals to HGP researchers") build "better mousetraps for the basic steps in molecular genetics," according to its chief operating officer. "This is speeding up the research clock. Instead of taking a couple of years, researchers are taking a month...."[12]

C. Thomas Caskey's remarks at a 1987 Office of Technology Assessment workshop on "issues of collaboration" are saturated with cravings for speed in the laboratory. After alluding to the "substantial errors" made by recombinant DNA researchers in "having developed this very powerful technology in the United States" and then having "immediately put the brakes on" in the 1970s, Caskey made it clear that he wanted to avoid any further sluggish impulses:

Every molecular biologist in the United States has known "clone by phone" for years. The communication and exchange between investigators in the United States has kept the research at an extremely high tempo...Professor X at Stanford has something that you've heard about at a Gordon Conference or a meeting and it would greatly facilitate your research in an area that is probably not identical to his own interest. You ring up the professor at Stanford, the next day in the mail by Federal Express comes the clone and you're off and running. You've cut back nine months, 12 months of developmental work and you're immediately moving on your project....

...I am definitely detecting, as I call my friends now and discuss these things, a tightening of this attitude. I would like to see us focus on ways of facilitating the exchange of information materials. And if we have cumbersome old patent laws that aren't staying up to speed with the speed of this technology and science, then let's rewrite it all, let's get it straight, let's maintain the leadership, let's facilitate these actions.[13]

This brief excerpt points to many other nodes of the genomics speed assemblage that require attention for a thorough understanding of what speed is, how it is achieved, and what it means – links to telecommunications technologies; dapperly uniformed Federal Express agents transporting styrofoam containers of carefully prepared biomaterials between what we still quaintly distinguish as "private corporate labs" and "academic" or "government" laboratories; even more dapperly uniformed attorneys transporting bits of text to create new ways to streamline "cumbersome old patent laws" or regulatory policies;[14] the relentless subdivision of research specialties into problems and trajectories "probably not identical" to those of the ever-present competitors; and that most deliciously vague allusion to a "tightening of attitudes" regarding the sharing of biomaterials and bio-information. Some of these topics will be touched on later, they would slow us down too much now.

It was in part the success of the new genomics technologies of the early and mid-1980s in producing speed that made something like the Human Genome Project see necessary. One indicator of this speed relationship can be found in the Human Gene Mapping Workshops, begun in 1973 as a way of compiling, organizing, and sharing results and information on the gene and genetic maps being built from the increasing data generated by new cytogenetic and molecular technologies. These roughly biannual workshops underwent rapid and dramatic change with the introduction of restriction fragment length polymorphism (RFLP) and other DNA-based technologies around 1980. Committees for specific chromosomes began proliferating, new committees on nomenclature and DNA methodologies were created.

The number of genes and cloned arbitrary DNA segments being mapped increased by as much as sixfold between 1981 and 1983, and the number of conference participants had doubled.[15] By the 1985 meeting in Helsinki, the total number of sites mapped (1,479) had doubled again, as had the number of cloned genes.[16] The most immediate and pressing need, however, was for a "system for permanent storage and updating of human gene mapping data." Each previous workshop had "acted on its own, thus unnecessarily duplicating earlier work, especially with regard to typing and tabulating".[17] Informatics was becoming the most important speed nexus.

The Committee on Human Gene Mapping by Recombinant DNA Techniques announced in 1985 that "a major restructuring of this committee's methods of data collection and analysis will be required so that compilation can be continuous between now and the next workshop." Data handling could not longer be done on "an ad hoc basis. Extensive databases will need to be created, maintained continuously, and reviewed at the workshops."[18] Computers had first been used at the Los Angeles workshop two years before, but they are used primarily as word processors. It wasn't until HGM9 in Paris in 1987 that the first "true database" was used, although that database could only be searched and not modified on-site. Only when the workshop arrived back at Yale for HGM10 in 1989 was an interactive, on-line database in use.[19] By that time, the workshops were said to be in "transition"; the nature of these changes in "a critical period" was best exemplified by the growing importance of computers. With what seemed to be a firm U.S. commitment to an organized mapping and sequencing project recently announced, "the single most problematic aspect" of such a project was to define and develop "new organizational structures to enable the international community to integrate information into the ever-changing map in a timely, accurate, and accessible form." "The task of monitoring and reporting construction of the human gene map," Ruddle and Kidd suggested, "has become a full-time job."[20]

In an early expression of a kind of add-fuel-to-the-fire philosophy that would be fully realized with the creation of the Human Genome Project, workshop organizers discussed some ways in which this already rapidly advancing specialty could be further accelerated. They argued for increased sharing and standardization of materials – libraries, somatic cell hybrid panels, lymphoblast cultures – as the way to achieve this. Even cloned probes, they suggested – the fundamental unit of mapping work, the basis for securing scientific credit and patentable products – should be exchanged, at least after a laboratory's "particular project" had been completed (Skolnick and Francke, 1982:194).[21] The solution to the problems of speed was to go faster.

3. Not So Fast...

This is one way to view the effects of the development of genomics. Time now to fiddle with the other analytic IV and inject an opposing dosage. During a recent interview with the head of a university-based molecular biology laboratory, my respondent flipped through the pages of the most recent issue of *Biotechniques* on his desk as he talked to me and answered my questions about commercially available kits:

One thing about a lot of these kits is that some just don't work – period. And some of them don't work within the context of the information, and the way you interpret those instructions. And I don't do troubleshooting for these kits. In fact, I have a long antagonistic history with a lot of these companies who put things out on the market with the expectation that if you have troubles, call us, and maybe we'll help you fix them or maybe we'll tell you what to buy next because we've figured out what the problem is. You go through that, there's 20 different plasmid purifications, 20 different this, 20 different that, and if my buy into that is considerable and it doesn't work, I can't afford that sort of stuff.

I make a choice that says that in a lot of cases, we will not buy kits where I don't feel they have a lot of obvious advantages to them. Unless somebody comes in to sincerely push that, or they manage to get me in a good mood. So here's this ad for a vector, which is used…it's just a vector. And we already have this vector, so we don't buy this, we just propagate it ourselves, which is probably illegal. Certain times we use some pieces of junk to do plasmid purifications, but we're really not a kit-dependent process. Largely because I have a lot of students here, and the problem in that sort of situation would be with the exception of spending a lot of time in management type mode, where you're coordinating activities, which is something I choose not to do. It's sort of like a little Darwinian experiment out there, and I just choose not to get involved in the micromanagement of the process. What that means is that there's like, you know, 15 different little franchises out there, and if you're going to satisfy everybody who wants a kit, they want 15 kits. So we largely don't buy into a lot of that. So you have to sort of look at those things and say, what is really there?[22]

When we came across an ad that had a stopwatch in it, and I confessed my liking for that particular image, he added sardonically, "Yeah, well, you're just sitting there, and it's like: I'm really concerned about doing this in 15 seconds, so I can drink more coffee. It's a bunch of shit."

Speed advertised is different than speed achieved. The IV with the uppers works only in conjunction with the IV with the downers. Machinic speeds only exist in collision and collusion with other speed regimes of tacitly skilled craftworkers, management styles, laboratory life, skirting licensing agreements, and economic relationships.[23] Regarding the latter, note how rapidly the language shifted between a metaphor of Darwinian struggle on the lab floor to one of 15 different little franchises. The intertwining of speed and the distribution of capital came up at another point in the interview, when we were discussing randomly amplified polymorphic DNA, otherwise known as RAPDs:

You have to understand that the technology, although it's available, is not universally available. I mean, sure, I would love to have a DNA sequencer. I would love to have my own DNA synthesizer. But it's a hundred thousand dollars, OK? I don't have a hundred thousand dollars. It's like – going back to this RAPDs – I was at

DuPont on Tuesday. DuPont is the land of RAPDs, and you know, we were sort of talking about thermocyclers. And there are different grades; it's just like cars. So we have sort of like the Yugos of thermocyclers. They work well, but they don't work as well as they should. And we accept that and we deal with that and we learn certain tricks and we get along OK. The sort of Cadillac [model] is a Perkin-Elmer 9600, and it has certain intrinsic benefits, but it's also three to five times more expensive than the ones that I have. So you go to DuPont, and the guy starts talking about, "oh yeah, we have, we have, uh"...the guy didn't even know how many 9600s they had![24]

Here again the language and metaphors change swiftly, this time calling upon that favorite speed machine of boys across America, the car. Not everyone has the vehicle of their dreams: some are born to it, some work hard for it, some tinker with what they have. And the line "I was at DuPont on Tuesday" is tossed off in a very matter of fact way, suggesting that the distinction between public and private science no longer does much work for the study of genomics and its associated biotechnologies. Rather, we have a more or less rapid oscillation between these two frozen poles: if it's Tuesday, it must be DuPont.

4. The Impetus of Speed in the Human Genome Project

The problems of speed and informatics were expressed in some of the first legislation that sought to institutionalize the Human Genome Project in the United States, the Biotechnology Competitiveness Act of 1987. The centerpiece of this bill was the establishment of the National Center for Biotechnology Information at the National Library of Medicine. The swell and rush of new sequence data being generated with new, standardized molecular genetic packages had by 1984 resulted in a 12-month delay between data publication and entry into GenBank, then in operation for two years.[25] GenBank administrators hired additional clerical workers, invented new arrangements with authors and journals, and by 1986, the time delay was down to six months, but further data acceleration was already foreseen with the then-hoped-for Human Genome Project.[26] As the 1987 bill stated, "Information on the map and sequence of the human genome, as well as the many other important fields of biotechnology research, is accumulating faster than can be reasonably assimilated by present methods," making it "essential that advances in information science and technology be made so that this vast new knowledge can be organized, stored, and used."[27] To organize, store, and use still more data, faster still.

Certainly, there were those who did not see the need for acceleration, especially if it meant new bureaucratic mechanisms and a shift in priorities that might slow down other areas of research. The remarks of Robert Martin and Bernard Davis, cited in my introduction appeared at a time of heightened criticism of the Human Genome Project, albeit criticism that came after the National Center for Human Genome Research had already been established within the NIH.[28] But Davis was nevertheless invited to testify before a U.S. Senate committee in 1990 to air his criticisms of the Human Genome Project. It had been redefined and renegotiated over the previous few years by a relatively small group of genomics proponents meeting under changing institutional labels – the National Academy of Sciences panel, or the NIH Ad Hoc Program Advisory Committee on Complex Genomes, for example. While Davis lauded the de-emphasizing of sequencing and the new stress on mapping, informatics, and the analysis of model organisms, he questioned whether this amounted to anything so innovative that it merited such special treatment as a separate NIH granting mechanism:

These are all excellent goals. But except for the mapping I would suggest that they are what we would be doing today if there were no human genome project. We would be doing all of these things through competing grants, though no doubt on a smaller scale. If this project were to be submitted today as something new, without the history of the original political appeal of the very definite goal of something like putting a man on the moon – getting that last human nucleotide – if it were to be proposed today as a new project, it seems doubtful that it would generate the same political appeal that the original one had.[29]

Davis alludes here to a persistent definitional problem that genomics proponents had faced: What was the difference between a Human Genome Project and business as usual, i.e., competing grants "on a smaller scale?" As we have seen, that "smaller scale" was already producing more data than was possible to manage effectively. Arguments for the Human Genome Project in the mid-1980s boiled down to speeding up what was already going on.

This is evident in Walter Gilbert's introductory remarks at the 1986 Cold Spring Harbor session, where the desire to completely map and sequence *the* human genome was first discussed among the larger community of molecular biologists. Gilbert laid out three possible time horizons:

If the current state of sequencing continued for the indefinite future, it would take something of the order of 3,000 years to sequence – something of the order of 1,000 years, 1,500 years, to sequence this amount of DNA.

Now, what can one do about this? One possibility is to say: fine, let us let the current way in which we do science continue, as far as this project is concerned; let's simply wait for the gradual improvement in techniques. So that's one choice. And that probably does mean that it won't take 1,000 years with the gradual improvement of techniques, but it does mean that it will take probably of the order of 100 years.[30]

The second possibility was to "focus some level of effort at the problem" over a "moderate period of time," roughly 20 or 30 years. And finally:

The third possibility is to try to do this in, let's say a scientist's immediate lifetime, a much shorter time than this, and let's say, let us try to develop a map of the human genome, and a sequence, by the year 2000. Fifteen years from now. I think the third is a more striking proposal. The third is a proposal in which, if you discuss what you could learn from looking at the human genome, it's something which you can imagine being here in a reasonable time.

"I doubt one can energize around" the first proposal, Gilbert reflected. Like many "moderate" proposals, Gilbert's second option faded quickly from consciousness as well. Gilbert and a few others would spend the next several years trying to convince the appropriate people, trying to "energize around" the idea that "a scientist's immediate lifetime" was indeed a "reasonable time."

5. "Full Speed Ahead" Arguments

A number of arguments were advanced in support of this high-speed option. Occasionally, they would be in the most abstract and general terms, as manifest in remarks made by Genentech CEO and President G. Kirk Raab at a congressional hearing in support of the Human Genome Project:

At Genentech…we support 100% of this effort, and I briefly want to tell you why.

First, speed. The speed which this will give us to solve problems can be likened to what has happened in the computer sciences, and what can do today, as compared to just five and ten years ago, the speed of science is tremendous, and the ability to use technicians rather than scientists to make things happen.[31]

Granted that government hearings are not always the occasion for specificities and direct remarks; still, it seems that in some ways speed has escaped any specific rational justifications – or rather, that there is something in addition to those rational justifications offered. That speed can be *built in*, standing in for the skills of the scientist at the reduced labor cost of the technician: one can readily understand the president of Genentech making such a calculative rationale. But there seems to be something more expressed here, something at once simpler and more complex, an apparent desire simply *to go fast*.

There is little doubt that formations of capital play an important part of the genomics speed assemblage and arguments for it, a point that comes out again in a number of ways later in this paper.[32] But by far the most frequent and effective speed argument had to do with the potential medical importance of genomics-based knowledge and practices. In late 1985, when the scientific and bureaucratic appetite for a large-scale, organized genomics project was still to be found only in the U.S. Department of Energy, Charles DeLisi E-mailed a colleague a few thoughts that presaged Gilbert's comments cited above. "With respect to the grind, grind, grind...argument," DeLisi replied to one of the anticipated criticisms that his colleague had voiced, "it seems to me that if the human genome is to be sequenced, there will be some grind work; what we are discussing is whether the grinding should be spread out over 30 years, say, or compressed into 10. I think a case can be made that the impact on medicine and biology will be so enormous relative to the costs of this project that acceleration is overwhelmingly desirable."[33]

Such arguments also served as a goad to continued rapid work once the project was under way, and meshed well with other speed incentives. "We have to get some real results in the next five years...find the gene for something which you might not have found if you didn't have the human genome mapped," as Watson said in regard to the need to keep congressional patrons interested and enthusiastic. But he quickly joined this political imperative to an ethical one: "If you want to understand Alzheimer's disease, then I'd say you better sequence chromosome 21 as fast as possible. And it's unethical and irresponsible *not* to do it as fast as possible."[34] Or as he put it to a Senate committee:

Some people say, well, why this targeted project, why the hurry? Why not just let the individual investigator (RO1) grants come in, and not have this big program. I think if you are a family with a serious genetic disease, you know, there are only a certain number of years you are going to live. You do not want that gene to be found 100 years after you are dead. So, if we can do this work over the next five years, I think we should do it...I think we are doing the moral thing to try hard to get these genetic resources...if you can find it now, I would rather have it this year instead of ten years from now. So I am in a hurry.[35]

To this linking of medical and moral imperatives, we should add personal motivations. I have already mentioned Walter Gilbert's urging at Cold Spring Harbor in 1986 to do this project "in a scientist's immediate lifetime," i.e., the lifetime of a scientist of his particular generation. Watson, too, has mused upon his own mortality in addition to that of the individuals from families with serious genetic diseases mentioned at the Senate hearings cited above, as well as the narrative demands of his life and career:

People ask why *I* want to get the human genome. Some suggest that the reason is that it would be a wonderful end to my career... 'That *is* a good story...' The younger scientists can work on their grants until they are bored and still get the genome before they die. But to me it is crucial that we get the human genome now rather than twenty years from now, because I might be dead then and I don't want to miss out on learning how life works.[36]

6. The Speed of Politics and the Politics of Speed

> "To govern is more than ever to fore-see, in other words to go faster, *to see before*."[37]

As Barney Glaser and Anselm Strauss have pointed out in their exemplary research on the temporal aspects of "status passages": "*The legitimator who must create a timing for a seldom tried, new, or temporally unknown passage has more discretion in setting the timing* than has a legitimator for a passage wherein the timing is generally well known, and typically scheduled and organized – e.g., a school year."[38] As the chief legitimator for the Human Genome Project, James Watson – whose own career was founded on the superior velocity exhibited in the "race" for the double helix – was instrumental in speeding up genomics by means of his advocacy for this enlarged, accelerated, and independent funding mechanism with the fancy name. Just as Gilbert's introductory intervention at the 1986 Cold Spring Harbor meeting pushed the "fast track" option, Watson, too, worked both the molecular genetics community, the halls of the U.S. Congress, and the official committees, which he helped establish and on which he served, to keep genomics moving apace. Without going into detail on the many instances in which Watson flexed his political muscles, I will mention just one episode from the National Academy of Sciences/National Research Council (NAS/NRC) committee's meetings, involving the level of funding and, tied to that, the speed at which events would proceed. Recall that at the Cold Spring Harbor meeting, Gilbert set out three possible futures, which varied in speed and commitment of resources; the fastest and largest proposal was made out to be the one that "energy could be organized around." A similar scenario was played out on the NAS/NRC committee, as described by Cook-Deegan:

A subgroup was delegated the task of producing budget options. Botstein spear-headed this effort and presented three options: $50 million, $100 million, and $200 million, with completion dates sooner for the higher figures (the year 2000 for $200 million versus 2025 for $50 million).... Watson objected to the process, noting that it would naturally incline the committee members to seem reasonable by choosing the middle option. He therefore suggested an option of $500 million-per-year. Since Botstein had already dubbed the $200 million annual budget the "crash program," Watson's became the "crash-crash."[39]

Watson appears as a savvy politician, well versed in the dynamics of group decision making, intervening at the appropriate time to keep the genomics vehicle not just moving, but *crashing*. After some discussion within the subgroup and a "review" of figures by Botstein, $200 million per year was the sum settled on by the committee to recommend.

Standardized genomics packages do not just produce speed but require also this kind and level of capital commitment.[40] Speed appears then to be at least partially the result of deliberate, debatable decisions. This externalized "crash" speed of funding level was also *internalized*, and expressed what demarcated the Human Genome Project from those things that Bernard Davis (cited earlier) argued would be done anyway. On the NAS/NRC panel, as Cook-Deegan has reported, yeast genomicist Maynard Olson argued that "projects should be considered genome research only if they promised to increase scale factors threefold to tenfold (size of DNA to be handled or mapped, degree of map resolution, speed, cost, accuracy, or other factors)."[41] The definition of Human Genome Project research revolved around acceleration. Olson's defini-tion was adopted by the NAS/NRC in its 1988 report, which was so important to establishing an image of community consensus on the Human Genome Project: "The human genome project should differ from present ongoing research inasmuch as the component subprojects should have the potential to improve by 5- to 10-fold increments the scale or efficiency of mapping, sequencing, analyzing, or interpreting the information in the human genome."[42]

We might also ask: Who stands in the best position to benefit from speed? What kinds of structural advantages might exist, or might emerge, or might be created? Defining the "Human Genome Project" in terms of speed resulted in a number of structural inequalities, or so some charged. Genomics proponents of a certain type and at a certain level of lab size and achievement – Maynard Olson, Leroy Hood, Walter Gilbert, C. Thomas Caskey – who had been instrumental in lobbying the project into existence, were also able to provide the new speed levels being requested. The project was defined largely in terms of what they were capable of and interested in doing.[43] Critics charged that too much money was going to large-center grants. After it was established, Watson wanted an even "faster and more accountable genome project" and advocated an even more pronounced emphasis on contracts and centers: "It's my belief that the program has got to change emphasis. To get the job done, we need contracts, not grants...."[44] To what extent does speed entail a certain level of cronyism, the strengthening and further stratification of speed hierarchies?

7. Gold Rushes, lambda-ZAP, BLASTX, FASTA, and the Spewing Business: Projecting Speed Genomics Fully Realized

Craig Venter's scheme, developed while working at the NIH, to pioneer the "business of spewing out sequences of gene pieces"[45] known as cDNAs and patenting those fragments of expressed genetic sequences as tools for further biomedical research, has received a great deal of attention from many concerned quarters.[46] Most of this attention has revolved around issues such as whether or not cDNAs are "really" "useful" and hence "really" patentable, or the deplorable fact that, according to some professional bioethicists (whose own importance develops in speedy tandem with genomics), "this is not science. This is like the Gold Rush"[47] – as if that wasn't precisely the point. Venter's scheme stands as an epitome of the genomics speed assemblage, combining standardized packages, fluid and abundant capital, and some degree of career building into a "land grab" (*ibid.*) of impressive proportions. "Venter's work highlights the speed of the Human Genome Project," stated the Association of Biotechnology Companies' president Forrest Anthony, and it was the unleashing of such pure speeds and the unregulated competitive struggles that they would lead to that disturbed the more moderate speed enthusiasts.[48] It was just too raw.

Even the origin of this plan, journalistically reconstructed in the favored Eureka! format, is situated in persistent speed vectors, metaphorical and actual." Frustrated by the snail's pace at which he was able to decipher the human genetic material," Venter struggled for years to solve the speed problem, when "on an airplane coming home from a meeting in Tokyo," inspiration hit.[49]

But the ephemeral vehicle of inspiration crashing instantaneously into Venter's brain had to be joined to the controlled high-speed wreckage of the standardized, automated genomics packages discussed earlier. "Three commercial human brain cDNA libraries" marketed by Stratagene as lambda-ZAP catalog numbers 936206, 936205, and 935205 were "converted en masse to pBluescript plasmids" (a Stratagene vector system); "Qiagen columns (Studio City, California) improved the percentage of plasmid templates that yielded usable sequences"; these templates were used in the PCR process, and the results of that speedy reproduction were run through the Applied Biosystems DNA sequencer. The resulting sequences were examined for similarities in GenBank, and their possible protein analogues were matched against the Protein Information Resource database. Those searches were conducted "with our modifications of the 'basic local alignment search tool' programs for nucleotide (BLASTN) and peptide (BLASTX) comparisons" – programs based on an algorithm that "sacrific[es] some sensitivity for the 60- to 80-fold increase in speed over other database-searching programs such as FASTA."[50]

All of which appears, however, to have been too fast for the National Institutes of Health to contain it. Venter's genomics vehicle ran off the federal rails at midnight, July 13, 1991, when he and 30 colleagues (nearly his entire staff) transferred their efforts to the new Institute for Genomic Research and its profit-making arm, Human Genome Sciences Inc., fueled by $70 million in venture capital, which their speed technologies had attracted. As the *New York Times* contextualized it:

Wallace Steinberg, chairman of the board of HealthCare Investment Corp., which is financing Dr. Venter, said he suddenly realized that there was an international race to lock up the human genome. If Americans do not participate, he said, they will forfeit the race and lose the rights to valuable genes to Britain, Japan and other countries that are in the race to win. He said the National Institutes of Health could not afford to invest enough money in Dr. Venter's enterprise to make it truly competitive.[51]

Steinberg's comments here echo almost exactly the arguments made for the Human Genome Project as a whole. It is Darwinism writ large, small, any scale you can imagine. For nations, corporations, genomicists, technoscientific tools, and biomaterials alike, it becomes a matter of survival of the fastest.

8. Concluding Lines of Flight

This essay began by fleeing from the category of "Big Science," and so it returns there before grinding to a halt. The Human Genome Project has been compared most frequently with two other technological projects: particle accelerators (especially the relatively contemporaneous "super-conducting supercollider") and the Apollo project. The comparisons are almost always cursory and undeveloped, rarely going beyond the level of "they're each tools," or "they're each exciting explorations," or, most frequently, "they're all examples of Big Science." Occasionally such analogies would be challenged on equally uninteresting grounds: a particle accelerator was *one* machine, located in one place and limited in its access and utility, and the Apollo project had *one* goal (i.e., the moon), and both had to be *completed* before they were "useful." By contrast, the Human Genome Project and its informational and technological fruits, the argument went, would be a resource used by many scientists in dis-parate fields and would be useful immediately, at every step of the way, to "completion."

But let's not be so hasty in dismissing these analogies, for we can get some mileage out of them. I am particularly interested in the localization issue. Particle accelerators/detectors are esoteric machines, a collection of rare materials tightly constrained and controlled, assembled by many construction workers, technicians in far-flung industries, and scientists for the purpose of producing phenomena that are otherwise unobservable. That is, they are out of the realm of general human knowledge, for an elite group of (primarily) men interested in so-called "fundamental" knowledge of the physical world. While it is true that they don't produce these phenomena until they are "completed," having been calibrated, standardized, and tinkered with, they are the locus of jobs and new knowledge associated with their various component systems during every stage of construction. The trick is to see the particle accelerator/detector not as a solitary technological apparatus, but as the central nexus in a vast, diverse, and spatially and temporally dispersed network of things and actions.

Similarly, it is a mistake to localize "the Apollo project" in the Saturn V rocket in the highly circumscribed earth-moon trajectory. As Walter McDougall and Dale Carter have shown so convincingly, the Apollo project was the equivalent of a technological and cultural *smear* across the U.S. social and temporal landscape, requiring an immense number of inputs in the form of personnel and devices, and performing a stunning array of jobs, from "promoting economic competitiveness" by propping up the aerospace/armaments industrial sectors (especially those in the Sunbelt), to solidifying a Cold War culture and the political fortunes of certain individuals within it.[52] The transforming effects of the Apollo project on the government-science network still reverberate in the time of the Human Genome Project:

In these years the fundamental relationship between the government and new technology changed as never before in history. No longer did state and society react to new tools and methods, adjusting, regulating, or encouraging their spontaneous development. Rather, states took upon themselves the primary responsibility for generating new technology. This has meant that to the extent revolutionary new technologies have profound second-order consequences in the domestic life of societies, by forcing new technologies, *all* governments have become revolutionary, whatever their reasons or ideological pretensions.[53]

Carter describes much the same territory, from the more productively imaginative perspective of Pynchonian paranoia, as the creation of the "Rocket State." From this perspective, the Apollo project can be seen as a localized effect of a much larger occurrence, as

the ultimate expression of both the Rocket State's joint command structure and of its uniformly seductive appeal: a celebration of liberty and security, of power and surrender, of escape and refuge produced through the combined efforts of those technical experts, businessmen, and propagandists called to the Rocket State's aid.[54]

If we subject the Human Genome Project to the same kind of diffracting gaze and see it dispersed as Projecting Speed Genomics, then the analogies hold up much better. Projecting Speed Genomics is not localizable, or rather it can only be localized in many places at once. As an NIH official phrased it in response to the criticism that the Human Genome Project was a "massive scientific assembly line": "The HGP is *not massive but distributed* in many labs across the country."[55] To talk about the genomics project is to talk about a much more amorphous set of activities, events, processes, and desires that has to do with the production and maintenance of a scientific infrastructure, that nineties buzzword. Its temporal boundaries are less distinct than the Human Genome Project's, extending back to the development of many of the basic recombinant DNA tools and techniques of the mid-1970s. It is a largely invisible/submerged/tacit network of standardized collective resources, which include biomaterials – cell lines developed in many laboratories and stored in places like the American Type Culture Collection, commercial restriction enzymes produced in a burgeoning biotechnology corporate service sector, DNA probes and gene libraries from university groups or the national labs, and virtual materials like the DNA sequences stored in GenBank and other databases. The genomics infrastructure also includes the standardized tools and embodied techniques that I mentioned earlier, with which those materials can be reconstructed into robust results and new configurations for further development by other laboratories, working both competitively and cooperatively in the increasingly indistinct worlds of universities, industry, and government.

Still, by moving from the noun "project" to the verb projecting, I hope to preserve some sense of individual and collective agency and hence responsibility in the midst of this dispersion. As some of the episodes that I described earlier illustrated, speed has been at least in some sense, in some instances, advocated and chosen deliberately. Producing the "fast track" has been hard work, with many people having to be convinced that a compressed time line was a matter of urgency and priority. But other episodes and texts suggest that such political choices have also been reactions to orders of technoscientific speed and their effects, exceeding the concept of human intention: the pervasiveness of highly mobile capital, intense careerism, relentless economic and nationalistic jockeying, sheer fascination with rapid innovation and speed – all of which have contributed to predictable but nevertheless surprising reworkings of the conceptions of health, class, privacy, and the body.[56]

My intent here has not been to install "speed" as a privileged explanatory category in science studies; "Fast Science" would quickly become as cumbersome, inexact, and procrustean a concept as "Big Science." The Human Genome Project does not "reduce" into speed; speed does not explain it all. But by imagining the Human Genome Project dispersing into Projecting Speed Genomics, form into speed body, nouns into verbs, speed becomes a useful trope for mapping the social, institutional, and conceptual reconfigurations being created in and through human genetics today. As Projecting Speed Genomics continues to create and re-create infrastructures, vehicles, and destinations all at once, we will have to respond with our own rapid reconceptualizations. If human genetics now is a vehicle – with an inscrutably alien and complex control panel – on a crash course, subject to a nonexistent map and itinerary, this work of reimagination will be a prerequisite for any possible actions other than the wide-eyed stare, the reflex move of the arms in front of the face, the scream before an impact that will never fully arrive but will always (be) becoming.

Notes and References

1. W.T. Lhamon, Jr., *Deliberate Speed: The Origins of a Cultural Style in the American 1950s* (Washington, DC: Smithsonian Institution Press, 1990).
2. Martin A. Lee and Bruce Shlain, *Acid Dreams. The Complete Social History of LSD: The CIA, the Sixties, and Beyond* (New York: Grove Weindenfeld, 1992 [1985]), p. 7.
3. Far from condoning these particular experiments or promoting a general valorization of the practice of spying, I am deploying the metaphor here for two effects: first, to recall another set of meanings of "to spy", troped less around secrecy and espionage and more around notions of "observing closely" and "keeping watch"; and second, as a reminder that these meanings of spying, like the meanings of genomics discussed in this essay, are being renegotiated under the changed circumstances of the post-Cold War world. On this latter point – especially as it has to do with the accelerated flows of detailed public information on natural resources, economic indicators, and so on, and the challenge these speeds present to the CIA analyst – see Herbert E. Meyer, "Reinventing the CIA," *Global Affairs* (Spring 1992), pp. 1–13. On the linkages between spying, terrorism, and speed, see James Der Derian, "Spy Versus Spy: The Intertextual Power of International Intrigue," in *International/ Intertextual Relations: Postmodern Readings of World Politics*, ed. James Der Derian and Michael J. Shapiro (Lexington, Mass: Lexington Books, 1989), pp. 163–187.
4. See Michael Fortun, "Mapping and Making Genes and Histories: The Genomics Project in the United States, 1980-1990," unpublished doctoral dissertation, History of Science, Department Harvard University, 1993.
5. Paul Virilo, *Speed and Politics*, trans. Mark Polizzotti (New York: Semiotext(e), 1986 [1977]), pp. 119–120.
6. Verena Andermatt Conley, "Eco-Subjects," in *Rethinking Technologies*, ed. V.A. Conley. (Minneapolis: University of Minnesota Press, 1993), p. 87.
7. Robert G. Martin, "Why Do the Human Genome Project?" *NIHAA (National Institutes of Health Alumni Association) Update 1*, Autumn 1989, pp. 4–5.
8. Bernard Davis, "What's the Big Hurry? – Thoughts on the Human Genome Project," *NIHAA (National Institutes of Health Alumni Association) Update 1*, Autumn 1989, pp. 6–7.
9. Quoted in Larry Casalino, "Decoding the Human Genome Project: An Interview With Evelyn Fox Keller," *Socialist Review 91/2* (1991), p. 114.
10. Virilio, *Speed and Politics*, p. 47.
11. See Joan Fujimura's elaborations of the oncogene "bandwagon" (with all its implicit speed metaphors) as constructed with the aid of the standardized packages of theory, instruments, and craft skills of molecular biology: "Constructing Doable Problems in Cancer Research: Articulating Alignment," *Social Studies of Science, 17* (1987), pp. 257–293; "The Molecular Biological Bandwagon in Cancer Research: Where Social Worlds Meet," *Social Problems, 35* (1988), pp. 261–283; "Crafting Science: Standardized Packages, Boundary Objects, and "Translation," in *Science as Practice and Culture*, Andrew Pickering, ed., (Chicago: University of Chicago Press, 1992), pp. 168–211; and *Crafting Science, Transforming Biology: The Case of Oncogene Research* (Cambridge: Harvard University Press, forthcoming).
12. This claim of Bios Laboratories' marketing director, Deborah Consiglio, also points to the kind of genre jumping that genomics publications are now made to perform: according to her, Bios's product line has been "cited in 38 publications since 1990,"

qualifying it as "the most mentioned product line in the short history of the Human Genome Project." The fact that "they're in the literature as bona fide products... makes the marketing considerably easier." Quoted in Fred Gebhart, "Bios Labs Aims to Be Top Supplier to Human Genome Project Researchers," *Genetic Engineering News*, June 1, 1992, p. 20.

13. Office of Technology Assessment, U.S. Congress, "Issues of Collaboration: Transcript of a Workshop held June 26, 1987," (Springfield, VA: National Technical Information Service, U.S. Department of Commerce, PB88-162797), pp. 113–115.

14. As a related note, one of Aaron Wildavsky's conclusions about the politics of genetic engineering is that "while efforts at regulatory control are likely to be tried, they are even more likely to fail," primarily because "the speed of discovery is so great that regulation cannot keep up with it. There are too many holes in too many dikes for a regulator to keep a finger in all or more of them." Aaron Wildavsky, "Public Policy," in *The Genetic Revolution: Scientific Prospects and Public Perceptions*, ed. Bernard D. Davis (Baltimore: Johns Hopkins University Press, 1991), p. 96.

15. See Mark Skolnick, H.F. Willard, and L.A. Menlove, "Report of the Committee on Human Gene Mapping by Recombinant DNA Techniques." *Cytogenetics and Cell Genetics, 37* (1983), p. 210; and Robert S. Sparkes, "Human Gene Mapping Workshop VII," *American Journal of Human Genetics, 35* (1981) p. 1334.

16. See Albert de la Chapelle, "The 1985 Human Gene Map and Human Gene Mapping in 1985," *Cytogenetics and Cell Genetics, 40* (1985) p. 1.

17. *Ibid.*, p. 6.

18. H.F. Willard et al., "Report of the Committee on Human Gene Mapping by Recombinant DNA Techniques," *Cytogenetics and Cell Genetics, 40* (1985), pp. 360, 363.

19. Frank Ruddle and Kenneth K. Kidd, "The Human Gene Mapping Workshops in Transition." *Cytogenetics and Cell Genetics, 51* (1989), pp. 1–2.

20. *Ibid.*

21. For all their foresight, however, the committee still thought in very modest – i.e., slow-terms as far as information storage and retrieval went. The results of mapping workshops, they suggested, "could be kept in the Gene Library computer system at Yale University and could be distributed with Dr. McKusick's *Human Gene Map Newsletter.*" See Mark Skolnick and U. Francke, "Report of the Committee on Human Gene Mapping by Recombinant DNA Techniques," *Cytogenetics and Cell Genetics, 32* (1982), p. 195.

22. Interview conducted by the author, Cornell University, October 1993.

23. For a precise account of the difficulties and annoying obstacles to speed presented by so-called "standardized" techniques and packages, see Kathleen Jordan and Michael Lynch, "The Sociology of a Genetic Engineering Technique: Ritual and Rationality in the Performance of the Plasmid Prep," in *The Right Tools for the Job: At Work in Twentieth Century Life Sciences,* eds. A.E. Clarke and J.H. Fujimura. (Princeton: Princeton University Press, 1992).

24. *Ibid.*

25. See G. Christopher Anderson, "Genome Database Booms As Journals Take the Hard Line," *The Scientist,* October 30, 1989, p. 4.

26. See *ibid.*; see also Christian Burks, "How Much Sequence Data the Databanks Will Be Processing in the Near Future," in *Biomolecular Data: A Resource in Transition*, ed. Rita Colwell (Oxford: Oxford University Press, 1989), pp. 17–26.

27. U.S. Congress, House. *Biotechnology Competitiveness Act of 1988.* 100th Congress, 2nd session, October 13, 1988, report 100–992, p. 1.

28. For further analysis of criticisms of the "Human Genome Project" in the United States; see Fortun, "Making and Mapping Genes and Histories". See also Robert Cook-Deegan, *The Gene Wars: Science, Politics, and the Human Genome* (New York: W.W. Norton and Co., 1994); and Daniel J. Kevles and Leroy Hood, "Reflections," in *The Code of Codes: Scientific and Social Issues in the Human Genome Project,* ed. Kevles and Hood (Cambridge: Harvard University Press, 1992), pp. 300–328.

29. U.S. Congress Senate, Subcommittee on Energy Research and Development, Committee on Energy and Natural Resources, *The Human Genome Project,* July 11, 1990, 101st Congress, 1st Session, S. Hrg. 101–894, p. 119.

30. Author's transcript of taped discussion at Cold Spring Harbor meeting, "The Molecular Biology of Homo Sapiens," June 1986; tape recording deposited at the Human Genome Archive, National Reference Center for Bioethics Literature, Georgetown University, Washington, DC; see also Fortun, "Making and Mapping Genes for Histories" for further analysis of excerpts from this important meeting in the historiography of "the Human Genome Project."

31. U.S. Senate, *The Human Genome Project,* p. 101.

32. Capital itself, as Gilles Deleuze and Felix Guattari note, is reorganizing itself into new speed regimes, wherein "at the complementary and dominant level of *integrated (or rather integrating) world capitalism,* a new smooth space is produced in which capital reaches its "absolute" speed, based on machinic components rather than the human component of labor.... The present day accelerated forms of the circulation of capital are making the distinctions between constant and variable capital, and even fixed and circulating capital, increasingly relative; the essential thing is instead the distinction between *striated capital* and *smooth capital,* and the way in which the former gives rise to the latter through complexes that cut across territories and States, and even the different types of States." Gilles Deleuze and Felix Guattari, *A Thousand Plateaus, Capitalism and Schizophrenia* (Minneapolis: University of Minnesota Press, 1987) p. 492. For a discussion of the uses and meanings of "striated" and "smooth" in their topological recasting of philosophy, see *ibid.,* pp. 474–500, pp. 363–374; see also Brian Massumi, *A User's Guide to Capitalism and Schizophrenia* (Cambridge, MA: MIT Press, 1992), p. 6.

33. E-mail communication between Charles DeLisi and David Smith, December 30–31, 1985. "DOE Policies" file, Box BCD7, Human Genome Archive, National Reference Center for Bioethics Literature, Georgetown University, Washington, DC; see also Fortun, "Making and Mapping Genes and Histories," pp. 88–91.

34. Quoted in Stephen S. Hall, "James Watson and the Search for Biology's 'Holy Grail'," *Smithsonian,* February 1990, p. 46.

35. U.S. Senate, *The Human Genome Project,* p. 33.

36. James D. Watson, "A Personal View of the Project," in *The Code of Codes: Scientific and Social Issues in the Human Genome Project,* ed. Daniel J. Kevles and Leroy Hood (Cambridge: Harvard University Press, 1992), pp. 164–165.

37. Paul Virilio, *Popular Defense and Ecological Struggle* (New York: Semiotext(e), 1990), p. 87.

38. Barney G. Glaser and Anselm L. Strauss, *Status Passage* (Chicago: Aldine, 1971), p. 39.

39. Cook-Deegan, *The Gene Wars,* p. 131.

40. It is important to see this not as a one-way relationship – i.e., money producing tools – but more as the kind of "cycle of credit" described long ago (!) by Bruno Latour and Steve Woolgar: Bruno Latour and Steve Woolgar in *Laboratory Life: The*

Construction of Scientific Facts (Princeton: Princeton University Press, 1979), pp. 187–233.

41. Cook-Deegan, *The Gene Wars*, pp. 130–131.
42. National Research Council, *Mapping and Sequencing the Human Genome* (Washington, DC: National Academy Press, 1988), pp. 2–3.
43. In the course of my fieldwork on the "Human Genome Project," on several occasions, I heard stories about so-and-so from such-and-such institution, who got a grant that had a low priority score but was funded because of his connections with someone on the advisory panels or study groups. Because such remarks are generally made "off the record," they don't get built into our knowledge about genomics. Earl Lane, a reporter on the genomics beat for the New York daily *Newsday*, began to detail how genomics proponents who had served on the important "advisory" and "evaluative" panels to the HGP "stand to receive millions of dollars over the planned fifteen-year life of the program." See Earl Lane, "The Funding Ruckus," *Newsday*, October 23, 1990, p. 21, Lane also heard several such stories and requested through the Freedom of Information Act the raw priority scores of applicants to the National Center for Human Genome Research, a request that was denied, appealed by Lane, and the appeal denied by the Assistant Secretary for Health on the basis that, contrary to Lane's claims, such scores were evaluative in nature and their release would violate the privacy of researchers. (Earl Lane, personal communication, March 18, 1992).
44. Jeffrey L. Fox, "Faster and More Accountable Genome Project," *Bio/Technology, 10,* February 1992, p. 120.
45. Gina Kolata, "Biologist's Speedy Gene Method Scares Peers But Gains Backer," *New York Times,* July 28, 1992, p. C1ff.
46. See Edmund L. Andrews, "U.S. Seeks Patent on Genetic Codes, Setting off Furor," *New York Times*, October 21, 1991, pp. A1ff; Robin Eisner, "Biotechnology Community Mixed on NIH's Gene-Patenting Efforts," *The Scientist*, December 9, 1991, pp.1ff.; Patrick D. Kelly, "Are Isolated Genes "Useful'?" *Bio/Technology 10*, January 1992, p. 52; Leslie Roberts, "OSTP to Wade into Gene Patent Quagmire," *Science, 254* (1991), pp. 1104–1105; Leslie Roberts, "NIH Gene Patents, Round Two," *Science, 255* (1992), pp. 912–913; Leslie Roberts, "Scientists Voice Their Opposition," *Science, 256* (1992), pp. 1273–1274; Leslie Roberts, "Rumors Fly Over Rejection of NIH Claims," *Science, 257* (1992), p. 1855; and Scott Veggeberg, "Controversy Mounts Over Gene Patenting Policy," *The Scientist*, April 27, 1992, pp. 1ff.
47. Kolata, "Biologist's Speedy Gene Method…," p. C10.
48. Quoted in Eisner, "Biotechnology Community Mixed…," p.10.
49. *Ibid.*
50. All quotes from Mark D. Adams et al., "Complementary DNA Sequencing: Expressed Sequence Tags and Human Genome Project," *Science, 252* (1991), pp. 1651–1656.
51. Quoted in Kolata, "Biologist's Speedy Gene Method", p. C1.
52. Walter A. McDougall, *…the Heavens and the Earth: A Political History of the Space Age* (New York: Basic Books, 1985); Dale Carter, *The Final Frontier: The Rise and Fall of the American Rocket State.* (London: Verso, 1988).
53. McDougall, *…the Heavens and the Earth*, p. 7.
54. Carter, *The Final Frontier*, p. 10.
55. Dr. Elke Jordan's response to BIONET letter of Michael Syvanen et al. (no date; sometime in spring of 1990), "Domenici Hearing 7/11/90" folder, Box BCD7,

Human Genome Archive, National Reference Center for Bioethics Literature, Georgetown University, Washington, DC.

56. One final example of the conflict of speed regimes: at the Human Genome I meeting in 1989, Watson discussed "the ethics thing," expressing his commitment to individual privacy but also suggesting, "I think we don't want people rushing around passing laws without a lot of serious discussion." (James D. Watson, "Organization: NIH," address delivered at Human Genome I, San Diego, California, October 2–4, 1989, author's transcript). Having hundreds of scientists "rushing around" redefining the body and the ethical and social relationship between bodies is unproblematic, while any acceleration of the more "social" (i.e, regulatory or legal) regimes of action is cause for extreme concern.

THE PRACTICES OF PRODUCING MEANING
IN BIOINFORMATICS

JOAN H. FUJIMURA

Department of Anthropology (and Program in History and Philosophy of Science),
Stanford University, Stanford, California

In a very real sense, molecular biology is all about sequences. First, it tries to reduce complex biochemical phenomena to interactions between defined sequences – either protein or polynucleotide, sometimes carbohydrates or lipids – then, it tries to provide physical pictures of how these sequences interact in space and time.[1]

Along with the explosive growth of biological data we are currently in the middle of a parallel technological revolution, the use and understanding of modern computers and the associated software. The opportunity thus exists to combine the potential of modern computer science, database management, and artificial intelligence in a major effort to organize the vast wealth of biological and clinical data. The time is right for three reasons: the amount of data is still manageable even in its current highly-fragmented form; important hardware and computer science tools have been greatly improved; and there have been recent fundamental advances in our comprehension of biology. This latter is particularly true at the molecular level where the information for nearly all higher structure and function is encoded. As noted earlier in the Models for Biomedical Research, A New Perspective: "We seem to be at a point in the history of biology where new generalizations and higher order biological laws are being approached but may be obscured by the simple mass of data." The organization of all biological experimental data coordinately within a structure incorporating our current understanding – the Matrix of Biological Knowledge – will provide the data and structure for the major advances foreseen in the years ahead.[2]

Mark it on your calendars now: midnight, Sept. 30, 2005. That, predicts Professor Victor McKusick, is the moment when a researcher somewhere in the world will transmit the block of information that completes the gene map in the computerized Genome Data base located here at Hopkins. With that researcher's findings entered in the GDB, mankind will have, for the first time, a complete and easily accessible guide to every gene on every chromosome carried in every cell of every human body.[3]

49

Michael Fortun and Everett Mendelsohn (eds.), The Practices of Human Genetics, 49–87
©1999 *Kluwer Academic Publishers. Printed in Great Britain.*

Introduction

Molecular genetic sequence databases are fast becoming the basic informational resources *in and with* which biological knowledge is currently being collectively constructed. Sequence information is the newest set of reagents in biology. Information represented in bytes in sequence databases is used to construct new representations and new meanings, which in turn have their own consequences.

Molecular genetic sequence information is the latest set of representations which, through computerized databases, lie within and connect many laboratories, scientific disciplines, and representations of phenomena. It is a distributed network of collective representations in terms of both construction and use. That is, this network links previously unrelated biological elements that belong to different domains of knowledge.

A principal tool that researchers use to attempt to make "meaning" of the sequences of letter "codes" in the databases is the concept of homology, a concept which has been called "the central concept for *all* of biology."[4] After cloning and sequencing a gene or after sequencing a protein of interest, researchers commonly search the databases, looking for matches with their DNA or amino acid sequence product. These matches are interpreted via the particular definitions of homology employed in that laboratory at that time. I use the various definitions of, and debates over, the concept of homology to examine the local construction of meaning using sequence information.

These debates are important because they highlight the processes through which the new bioinformatics is creating new images of nature and humanity. These images are in part expressed in what scientists call a new "language" of DNA and amino acid sequences. The "vocabulary" of this "language" is constructed from "homologous" sequences, that is, similar sequences that share a common ancestry.[5]

This paper uses sequence information and homology construction to explore two problems. First is the problem of representation in molecular biology. What practices constitute biological representations? Through which practices do representations such as DNA and protein (amino acid) sequences become players beyond the local, contextualized situations in scientific laboratories? How do they become "codes" in the larger arenas of biopolitically universalizing definitions of nature and life?

Some practices produce representations that gain the authority of universality via constructions of continuity between knowledge production sites. In this paper, I examine several practices and forms of standardization that connect laboratories across the biological subdisciplines, and more recently and in a different way, across the disciplines of biology, statistics, mathematics, computer science, and engineering. These standardizing practices and technologies include molecular genetic sequence information, homology search algorithms, and computer databases and software. They create very particular representations of nature. Yet these particularistic constructions, taken as universal representations because of the standardized technologies, are being used to create future conceptions and realities of nature, humanity, life, health, and disease.

Molecular Biology as Information Science?

The Human Genome Project (HGP), largely institutionalized in the National Institutes of Health (NIH) and the Department of Energy (DOE) in the United States, has as one of its stated goals to map and sequence "the" entire human genome, projected to cost up to three billion dollars.[6] The human genome refers to the 23 pairs of chromosomes (an estimated 3 billion base pairs of DNA) that are a part of every human cell. The HGP has become a convenient focal point for discussions about human genetics and its medical applications as they have developed over the last decade, and it will continue to develop in the years to come. Prominent figures in molecular biology have argued for the importance of the HGP as a foundation for progress in biology and medicine.

As an effort to both promote the Human Genome Project and to provide frameworks for dealing with its consequences, some biologists have recently advocated for the transformation of molecular biology into information-based theoretical science.[7] For example, molecular biologist Walter Gilbert has promoted this transformation as the next step in the "progressive development" of molecular biology and as the answer to the much maligned technologization of the field. In 1991, Gilbert published an editorial entitled "Molecular Biology is Dead – Long Live Molecular Biology, A Paradigm Shift in Biology" in the "News and Views" section of *Nature*.[8] The article provoked much discussion, supportive and critical, among the biology and information science researchers I was then interviewing.

While several kinds of genomic maps are being created, the so-called "ultimate" physical map or the map with the highest degree of resolution, are the DNA sequences themselves. Sequencing genes is the reading out of the nucleotide bases that constitute the genes along the entire genome or parts of it. Gilbert's editorial argued that sequence information, stored in computerized databases and manipulated by algorithms packaged in software, are creating a paradigm shift in biology. He claimed that the increasing amount of collected sequence information is moving biology from an experimentally based discipline to a theoretically based discipline.[9]

In the current paradigm, the attack on the problems of biology is viewed as being solely experimental. The "correct" approach is to identify a gene by some direct experimental procedure – determined by some property of its product or otherwise related to its phenotype – to clone it, to sequence it, to make its product and to continue to work experimentally so as to seek an understanding of its function.

The new paradigm, now emerging, is that all the "genes" will be known (in the sense of being resident in databases available electronically), and that the starting point of a biological investigation will be theoretical. An individual scientist will begin with a theoretical conjecture, only then turning to experiment to follow or test that hypothesis. The actual biology will continue to be done as "small science" – depending on individual insight and inspiration to produce new knowledge – but the reagents that the scientist uses will include a knowledge of the primary sequence of the organism, together with a list of all previous deductions from that sequence.

According to Gilbert, as the genetic information produced and stored in databases grows, this new "theoretical biology" based on computerized sequence comparisons, manipulations, and simulations will also grow. Indeed, Gilbert and others have argued that there is no other way to make sense of the vast amount of information being produced. (Readers will note the bootstrap strategy by which Gilbert promotes the project and the transformation of molecular biology. He first advocates the initiation of the project; then he uses its outcomes to propel a major transformation in the style of biological research.)

How quickly will this happen? It is happening today: the databases now contain enough information to affect the interpretations of almost every sequence. If a new sequence has no match in the databases as they are, a week later a still new sequence will match it. For 15 years, the DNA databases have grown by 60 per cent a year, a factor of ten every five years. The human genome project will continue and accelerate this rate of increase. Thus I expect that sequence data for all of the model organisms and half of the total knowledge of the human organism will be available in five to seven years, and all of it by the end of the decade.[10]

Theoretical sequence analyses of various kinds are important in this context, since many of the relevant patterns are not immediately obvious. The sheer amount of sequence data now available also makes the appearance of a new breed of "theoretical molecular biologists" unavoidable, and if current proposals to sequence "the" human genome (as if this were a well-defined entity) are put into practice, automated data handling and, above all, data analysis must be given a very high priority.[11]

Shuffling Toward Theory

One example of such theoretical work is Gilbert and his colleagues' research on genetic evolution.[12] Using DNA sequence information, they proposed a theory arguing that "eukaryotic gene evolution proceeds through exon shuffling as well as through intron loss, intron sliding, and, possibly, intron insertion.[13] According to current molecular biology, exons are sequences of DNA, which code for messenger RNAs, which in turn code for proteins. Proteins are the active elements in cellular functioning. Walter Gilbert and his colleagues have proposed that "the segments of proteins encoded by the individual exons arose during evolution as small protein units [or modules] capable of independent folding and that they have assembled into multifunctional proteins as independent domains."[14]

Again, using computerized information-generated models, Robert Dorit, Lloyd Schoenbach, and Walter Gilbert proposed in 1991 that all the proteins currently existing in the world are descendants of 1,000 to 7,000 exons, which have been shuffled into different exon modules (mosaics) at different times through evolution in different organisms[15]. They argued that variation is produced through exon shuffling, and these variations might then be selected for or against in evolutionary processes[16].

Dorit and his colleagues further argued that modules of exons are separated by introns (like railroad cars separated by couplers). Introns apparently code for messenger RNA, but they do not produce a protein product. (In molecular biological language, introns are transcribed but not translated.) The purpose(s), or lack of purpose, of introns is debated. Researchers have not decided whether some or all introns have functions. Speculations are that introns form about 95% of the human genome. They have been called "junk DNA" partly because their activities are not yet understood. One theory is that the purpose of introns is to allow exon shuffling, because without these separators shuffling would be impossible.[18]

Speeding Toward the Information Age

The "paradigm shift" and Gilbert's efforts to promote it are not without
resistance, as one of my respondents noted. "Jones" is computer molecular biologist (originally trained as a physicist) whose paper was rejected
by experimental biologists for its lack of evidence. "Jones" coauthored a
paper that argued for a functional relationship between two proteins on
the basis of some computer work and "thinking". The paper was rejected
by a biology journal's reviewers on the grounds that there was no
experimental "wet lab" work to support the "speculation".

Furthermore, a common response of experimental biologists to my
question about Gilbert's proclaimed "paradigm shift" was "Not everyone
agrees with Wally Gilbert." Arguing against Gilbert's representations of
current experimental biological work, another respondent said, "Molecular
biology still requires thinking that goes beyond mere technical procedures." A biochemist whose work has gradually shifted to bioinformatics
said that he has also experienced resistance to his idea that there is a
"natural synergism" between molecular biology and information science.
He had submitted an abstract proclaiming this theme to a conference
supported by the NSF. His paper was rejected, and he guessed that it was
the resistance to this proposed transformation of molecular biology to
information science that his colleagues were rejecting. Indeed, his proposal read like Walter Gilbert's paradigm shift and quoted Gilbert's 1991
proclamation. In other cases, information-inclined academic biologists
have had trouble gaining tenure and promotion. Resistance was somewhat less prominent in industrial biotechnology companies.

Despite these forms and instances of resistance, computerized data
storage and analysis have received very high priority since Gilbert's writings.
Bioinformatics has already become a growing field, which includes the
efforts of computer scientists, statisticians/mathematicians, and engineers
in addition to biophysicists, molecular biologists, geneticists, and
biochemists. The *Annual Report I–FY 1990* of the National Center for
Human Genome Research (NCHGR) stated that "meeting the challenges
presented by genome informatics is complicated by the small number of
investigators who are available to work in the area. To reach its goals, the
NCHGR decided "to stimulate new collaborations among geneticists,
molecular biologists, computer scientists, statisticians and other mathe-
maticians, as well as experts in specialized integrated circuit design,
robotics, and other fields of hardware and software engineering" through
short courses, workshops, and interdisciplinary training programs.[18] The

reorganization of biological work had already begun and would speed up in the near future. Bioinformatics is now a prominent name used by various researchers, organizations, and professional groups. The International Conference on Intelligent Systems for Molecular Biology convened its third annual meeting in the summer of 1995, and the fourth in Japan in 1996. The Monbusho (Ministry of Education) in Japan is devoting an even greater proportion of its human genome project resources to bioinformatics over experimental biology than is the United States.

All these efforts are not aimed at Gilbert's, Morowitz's, and von Heijne's theoretical biology. Some are meant to assist molecular biologists' efforts to map and sequence "the" human genome in more efficient ways. However, as the human genome projects in the United States, Europe, and Japan contribute to the stores of sequence and map information in databases, there will be less need for that kind of work and more effort will be aimed at understanding the information. The mapping and sequencing may not be completed by the dates promised by the HGP proponents, but information is accumulating at an accelerating pace.

While sequence data is currently combined with genetic map data, results of specific experiments, and knowledge of organismal biological systems in order to make sense of the data, this situation will certainly change as more work is done using information bits/bytes as reagents. There is already more than enough information available for comparative analysis. Molecular biologists and evolutionary biologists are increasingly becoming quasi-informaticians, quasi-computer scientists, and quasi-mathematicians/statisticians in order to be able to "play" with the information databases.[19] Not only will they have to be able to manipulate and manage information in computers, they will also have to construct meaning using statistical methods (not experimental methods) as the amount of available data increase during the progress of the human genome project.

Once the [human genome] map and sequence data have been obtained, extracting useful information from them efficiently and economically will also require automated means. Meaningful patterns in the data can often be revealed only by statistical analysis, at the cost of many repetitive calculations. As the project continues and the quantity of data available for comparison increases, the size of even the simplest comparative analysis will exceed the time available to the human analyst.[20]

Even now, much work is oriented toward making meaning of the genomes through the use of statistics and mathematics. Consider the complaints of graduate students as they spend hours and days working on experimental procedures with all their inherent glitches. Consider the time and other resources that are saved when conducting virtual experiments using the computer instead (assuming that the computer systems are already in place). We know that the general speedup in research time has affected every discipline in favor of studies that are faster to complete, as well as in favor of disciplines that can produce more articles per unit time than disciplines whose articles and books require more investigative and write-up time.[21] Gilbert's vision may well take hold, and experimental biology might eventually become the work of technicians, as bioinformatics becomes the real work of molecular biology.

Information and Databases

When one "logs" onto a sequence database, one sees a linear readout of A, C, T, Gs for DNA; A, C, U, Gs for RNA; and A, R, N, etc, for the single letter codes for amino acids, along with information about the researchers, their publications, and other related sequences.[22] Databases are the computerized version of *publications* of sequence information, often accompanied by some other biological information. Before more efficient retrieval software programs for accessing the computerized databases were constructed, scientific journals and books published sequence information related to particular topics.

Since the early 1980s, searching the computerized databases has become the *routine* in all laboratories where gene or protein sequencing occurs and where computer facilities are available.[23] Cloning and sequencing a gene or protein has little meaning to molecular biologists unless that sequence information can be related to other kinds of information and other kinds of work arrangements. The information can be in the form of data about biological functions possibly related to the gene, which in turn requires follow-up work, an understanding of a biological system, and an in-depth understanding of the organism. Another option is to find matches for one's sequence and begin to construct a "history" for the sequence on the basis of the match(es).[24]

Because of the volume of information that has and will continue to be produced by the work of mapping and sequencing human genes, molecular genetic sequence information and genetic map databases have become a major focus of interest and concern to proponents of and participants in the Human Genome Project. The HGP has proliferated databases and increased the load on existing databases. Beginning with considerable support from the Department of Energy (DOE), databases of DNA, RNA, and protein sequence information about many different organisms have existed for some time. For example, two years before the HGI was lobbied into existence, GenBank (at Los Alamos National Labs) contained approximately 18 million base pairs of total DNA sequence information and approximately 2 million base pairs of human DNA sequence information.[25] In addition to human genetic sequences, DNA, RNA, and protein sequence information is also available for other organisms, including the much studied yeast, mice, *Drosophila*, *E. coli*, and the nematode *C. elegans*.[26]

Sequence databases allow scientists a faster and more efficient method for accessing information used in their experiments or for interpreting experiments. Researchers access the data through magnetic tapes, disks, CD-ROM[27], and direct computer-to-computer and computer-to-terminal transfer over telephone lines.

Databases are an important subject for social analysis because they contain the basic informational resources *in and with* which biological and biomedical knowledge is currently being collectively constructed. Information represented in bytes in sequence databases are used to construct meaning. Moreover, information in these databases is more equal than other representations because they lie within and connect many laboratories and other production situations. Databases are used by different laboratories and lines of research not only to share information quickly and easily but also to construct new meanings through information comparison. Collectively, these constructions will be influential in determining what it means to be human.

Some computational resources used in making meaning include software packages of sequence-analysis programs, which are currently being designed in both private industry as well as in university laboratories and institutes. Some of the analyses that scientists can perform using the database system include translation and location of potential protein coding regions; inter- and intra-sequence homology searchers; inter- and intrasequence dyad symmetry searches; analysis of codon frequency, base composition, and dinucleotide frequency; location of AT- or GC-rich regions; mapping of restriction enzyme sites; protein and nucleic acid structure prediction; signal-to-symbol transformations; and neural network models.

Using Homologies to Make Meaning of Sequence Information

There are many complexities, debates, and confusions encoded in this section's title. First, I will describe what homology searches are then I will move on to discuss these encoded complexities.

After cloning and sequencing a gene or after sequencing a protein of interest, researchers commonly search the databases looking for "matches" with their DNA or amino acid sequence product. These matches are interpreted via a particular definition of homology to construct some hypothesis about the sequence's function.

If you have done nothing else in terms of theoretical sequence analysis, no doubt you have more than once asked a colleague to run your latest sequence through a homology-search program in the hope that it will turn out to be related to some interesting protein already stored in one of the data banks. And your colleagues will more likely than not come back with a list of more or less surprising, more or less convincing "matches," a stack of dot-matrix comparisons, and one or two alignments of the best candidates.[28]

An early example of this routine search for homologies and its success is in oncogene research. According to Russell Doolittle, who was one of the first biologists to begin working with computerized sequence information, of the five possible outcomes of a homology search, "The Best' result is "Your sequence is obviously similar to something glamorous, like an oncogene."[29] In their efforts to promote the HGP, biologists Walter Gilbert and James Watson held up for display the case of oncogene research and praised the speed and efficiency with which oncogenes and genes previously thought to be unrelated were "found to be related" through computerized sequence matches. In various public speeches and articles in scientific and public media, Watson and Gilbert argued that the HGP would increase their abilities to find more homologies, which could then be used to better guide research on various diseases.

The following short example of a homology search from the early 1980s indicates the speed and efficiency valued by researchers and echoed in the rhetorics used as incentives for the Human Genome Project.

Several years ago, Joseph Brown ... purified a melanocyte tumor cell antigen. There was only enough material available for a single microsequencer run, and that endeavour only managed 13 cycles. In fact, only 10 of those first 13 amino acids were identified with any confidence. Still a search of our data bank turned up only a single candidate: transferrin ..., Brown promptly tested to see if the tumor antigen bound [59]Fe. It did, and with the same avidity as transferrin.... Today that protein, which appears to be an important factor in rendering the tumor melanocyte immortal, is known to be about 40% identical with the better known transferrins and is called melanotransferrin.... The take-home message is: A search of even a very short sequence may put you on the right track and save years of work.[30]

Software tools for performing sequence analyses include global sequence alignment, multiple sequence alignment, pattern matching, and identification of promoter regions. By streamlining the procedures and knowledge requirements for identifying sequence homologies, the computerized sequence databases allow undergraduate students not thoroughly trained in molecular biology, and certainly not trained in any particular "biological system", to do the work of searching for homologues.

Semantic Complexities and Practical Uncertainties of "Homology"

The aggressive confidence of modern biomedical science implies that we know what we are talking about. But a deeper reflection shows that this confidence is based more on hope than on certainty.[31]

I want to step back now to discuss *homology*, since it is a term one hears constantly around databases. As I said in the introduction, homology has been called "the central concept for *all* of biology."[32] Homology is an interpretive resource shared by biologists. It is a tool for making meaning of the sequences developed in different laboratories. However, like most cultural resources shared by many people, homology does not mean the same thing across laboratories. Homology does not consist of a stable, trans-situational body of rules, statistical measures, or ideologies. Neither is it made up from scratch every time it is used by scientists. The concept has a history, indeed several histories. I examine the concept of homology as used by different laboratories and fields.

In molecular-biology-laboratory shoptalk, terms like "homology," "similarity," and "identity" are not clearly defined. After hearing the word used in different ways in my field research among molecular biologists, I began to ask questions about the concept. When I asked for clarification, I was told that a homology is a similarity relationship, which also has some kind of evolutionary meaning. However, my search for further specification of the term led to diversity, debate, and controversy.

Probably no word causes more confusion in this field than the word "homologous". When two sequences are homologous, they share common ancestry. In this sense, there are no degrees of homology. Sequences are either homologous or they are not. Many investigators use the word when they mean "similar". Two sequences may be *similar* by chance, for example. They may *resemble* each other to a high degree, but they ought to be *very* homologous or *slightly* homologous. Also, two sequences may be 60% identical, but they are *not* 60% homologous.[33]

People in the field shudder when the terms "similarity" and "homology" are used indiscriminately: Similarity simply means that sequences are in some sense similar and has no evolutionary connotations, whereas homology refers to evolutionarily related sequences stemming from a common ancestor.[34]

[M]olecular biologists may have done more to confound the meaning of the term homology than have any other group of scientists...[T]he word homology is now used in molecular biology to describe everything from simple similarity (whatever its cause) to common ancestry (no matter how dissimilar the structures).[35]

The concept of homology has its origins in nineteenth century studies of anatomy, and it has gained greater saliency through the development of theories of evolution. Evolutionary biologists used the word "homology" to refer to comparisons of morphological forms, such as organs, between animals and species. Similar or "homologous" morphological forms were used to imply close evolutionary relationships and to build evolutionary genealogies or "trees". However, evolutionary biologists attributed diverse meanings to the term "homology", and debates ensued about those meanings among evolutionary biologists, as well as among historians and philosophers of evolutionary biology.

Homology is widely considered to be among the most important principles in comparative biology... As Colin Patterson put it, "all useful comparisons in biology depend on the relation of homology."[36] Yet there are still significant differences in the meaning attributed to the term (and related words such as analogy and homoplasy), even among authors professing the same general outlook. Advances in phylogenetic systematics and in developmental molecular biology have renewed interest in homology...[37]

Homologous (that is, similar) structures were *used as evidence for* evolution and phylogenetic relationships by Darwin and his contemporaries. Later, however, the existence of a shared common ancestry, that is, of a phylogenetic relationship, *became part of the definition of* homology; that is, "homology" now means "biological similarity that can be traced to common ancestry."[38]

Although ancestry was at first viewed only as an explanation for homology, it soon was incorporated into the definition... E.R. Lankester's paper of 1870 played an important role, as evidenced in later editions of the *Origin*. He was critical of essential similarity and instead connected homology and related terms to common ancestry. Furthermore, he suggested that homology be subdivided. For homologous similarity due to inheritance from a common ancestor, Lankester coined the term "homogeny, and for such similarity resulting from independent evolution he introduced "homoplasy." "Analogy" then referred to similarities that would not be accepted as homologous based on standard criteria. The term "homogeny" was never widely adopted, and "homology" quickly became associated with similarity that could be traced to a common ancestor. This kind of transition – from explanation to definition – is probably commonplace....[39]

This transformation of explanation to definition, of similarity as a consequence of shared common ancestry to similarity as meaning shared common ancestry, created a "semantic confusion" in the term "homology".

In retrospect, the [assumed] connection between homology and evolution created at least two difficulties that appear to be common sources of semantic confusion (Donoghue 1985). First, two potentially dissociable elements were combined in one definition.... If there are cases (as we suppose there are) in which similar structures originate independently, or cases in which very different structures originate through evolutionary transformation, one is forced to choose which element – similarity or ancestry – is to be given primacy. Second, homology was now expected to account simultaneously for the maintenance of similarity and for the transformation of form – for both constancy and change.[40]

In this paper, I can only give a few examples of the diverse meanings attributed to, and debates about, homology among evolutionary biologists and historians and philosophers of evolutionary biology. Much more could be and has been said on the topic.[41] My purposes for raising the historically and disciplinarily embedded semantic confusion in the definition and use of the term is that it has crucial implications for current uses of the concept in molecular biology and molecular genetics. Evolutionary biology and molecular biology are not discrete domains with discrete terms and meanings. Although "homology" originated with reference to comparisons of morphological forms such as organs between animals and species, the term presently encompasses other kinds of forms, including DNA sequences. Just as morphological forms have been used in building evolutionary genealogies, similarities among DNA and protein sequences are also currently being used to argue for homologous or evolutionary relationships and, in turn, for functional similarities.

A basic assumption in this use of sequence homologies is that genes contain "information" that is transferable and consequential for other events in an organism's functioning, development, and reproduction. This assumption of transfer of "information" underlies the idea that "meaning" can be made of the sequences of letter "codes" in the databases. The difficulty arises when one questions what "information" is. Do gene sequences "code" directly for phenotype? Is there information in the genes that prescribes development and function in a linear and direct fashion? Is this information complete or partial? If it is partial, what does it mean to say that genes "code" for phenotype?

The first problem presented to sequence analysis by this equation of structural similarity with close evolutionary relationship and therefore functional relationship in the term "homology" – that is, this movement from evidence to definition – is that evolutionary biologists object to it. There is plenty of resistance to the view that gene sequences "code" directly for phenotype and therefore that genetic similarities translate into phenotypic similarities.[42] Here is just one example, from V. Louise Roth. I discuss this issue more generally later in this paper.

The notion that homology is a relationship written in the genes is a tempting one, given that the genetic material does contain in coded form at least part of the information needed to generate phenotypic traits.... [However, s]imilarity in genetic sequence is no guarantee to similarity in phenotypic traits, due to the intervening complexities of epigenetic interactions and regulatory genes and pathways. In addition, quantitative genetic analyses in combination with selection experiments demonstrate that similar phenotypic end-products of selection for a specific trait (such as tail length) may be produced by different evolutionary pathways, resulting in different patterns of genetic correlation among traits in replicate lines...[43]

Roth's solution to this dilemma is to employ Van Valen's expanded definition of homology, which refers to the "continuity of information"

Van Valen ... proposed that homology is 'correspondence caused by a continuity of information.' In this context 'information' describes the genetic, epigenetic, and other mechanisms that specify phenotypes. The manner in which 'information' is transmitted defines the types of comparisons that are needed. The central issue in causal analysis of homology is recognition of the classes of biological processes that cause the invariance, lability, and individuality of morphological structures.[44]

Roth argues that "what is needed is not different definitions [of homology], but rather different explanations."[45] This "definition can be used by adherents to any school of thought by simply specifying the relevant kind of information."[46] However, George V. Lauder objects that "this very flexibility means that it is virtually impossible to apply the definition. How are we to judge the information content of two bones in different species? Might this not be especially difficult if we accept that these two [phenotypically similar] bones could have been produced by different developmental processes or even by genes... with differing patterns of genetic covariance and pleiotropic effect?"[47] A corollary complication in the transfer-of-information problem then is that similar phenotypes can be produced by very different genotypes, due, for example, to similar environments.[48] The critique then is that similar sequences do not necessarily produce similar phenotypes.

A second argument against the use of similar sequences to establish homology is that strong similarity between sequences can themselves be *due to* things other than common ancestry. Alternative explanations include random coincidence, analogy, reversal, convergence, or parallelism. Convergence happens when two sequences, or their biological functions, can independently evolve to become more alike or converge through time possibly as a consequence of selection. For some authors, analogy results from convergence and indicates a similarity that would not be accepted as homologous.[49] It is difficult then to establish common ancestry from similar sequences.

Even if one accepts the premise that homology is similarity, a third problem arises. How do we decide on criteria and rules for determining similarity? How similar must DNA or protein (amino acid) sequences be to justify the designation of homology? Although there is no final statistical test for deciding that a similarity is strong enough to warrant the label homology, some guidelines have been accepted and are being used by some biologists. For example, some molecular evolutionary biologists who use molecular protein sequence comparisons assume that 25% similarity for short sequences is not enough to conclude that the similarity is a homology. A 25% similarity could be considered adequate for making a claim of homology for long sequences, however. These biologists assume that their abilities to make correct decisions about homology increase with higher rates of similarity and longer stretches of protein sequences[50].

These rules are not hard and fast, nor are they universally practiced. Researchers often construct different cutoff rules for deciding on what percent similarities qualify as homologies. For example, in constructing their story of variation produced through exon shuffling and selected for (or against) in evolution processes, Dorit et al. used cutoff points that depended on the length of the sequences, as well as on the particular database used[51]. But other researchers have quarreled with those cutoff points.

The equation of similarity with homology [here used to mean common ancestry], and consequently, the acceptance of partial homology, is also widespread among molecular biologists... , despite arguments against such usage... and at least one journal that specifically prohibits it... (*Molecular Biology and Evolution*). Here, when it is said that two polypeptides or segments of DNA are 75% homologous, it means that they can be aligned in such a way that the same amino acids or nucleotides are present at 75% of the sites. In this context, the issue is complicated by the argument that sequences which are sufficiently similar must also be homologous in the evolutionary sense, simply because the chances of independently acquiring a high percentage similarity are vanishingly small....[52]

As the writer indicates, there is a profusion of disagreements and debates about the quantitative and qualitative guides for establishing homology from similarity among molecular biologists and evolutionary biologists. There is even more criticism of the *equation* of similarity and homology in molecular sequence analysis. Russell Doolittle and ten colleagues have criticized what they considered to be the sloppy and incorrect use of the term "homology" in many scientific articles. In a letter to the journal *Cell*, they recommended that molecular biologists confine themselves to the term "similarity" rather than "homology."

Similarity is relatively straightforward to document. In comparing sequences, a similarity can take the form of a numerical score (% amino acid or nucleotide positional identity, in the simplest approach) or of a probability associated with such a score.... A similarity, then, can become a fully documented, simple fact. On the other hand, *a common evolutionary origin must usually remain a hypothesis*, supported by a set of arguments that might include sequence or three-dimensional similarity.... We can deceive ourselves into thinking we have proved something substantial (evolutionary homology) when, in actuality, we have merely established a simple fact (a similarity, mislabeled as homology). Homology among similar structures is a hypothesis that may be correct or mistaken, but a similarity itself is a fact, however it is interpreted.... Evidence for evolutionary homology should be explicitly laid, making it clear that the *proposed* relationship is based on the level of observed similarity, the statistical significance of the similarity, and possibly other lines of reasoning.[53]

A reason for the vehemence of the letter is that while researchers often include qualifiers about statistical significance in their writings, they ignore their own qualifiers in doing their work. That is, while researchers might admit that "estimating the significance of an observed similarity score is still somewhat of a black art – [that] no satisfactory statistical theory that takes care both of the properties of the different alignment algorithms and the statistical properties of unrelated nucleotides sequences... exists," they continue to use statistical similarity scores to "confirm" homologies, that is, common ancestry.[54]

Some researchers have argued that homology can only be determined by other kinds of evidence, often that which is not available to molecular biologists. For example, Doolittle promotes the development of methods "for improving the sensitivity of sequence comparison schemes so that 'genuine' ancestral relationships can be distinguished from chance resemblances."[55] These include comparing ancestral sequences, multiple comparison methods (using more than two sequences), comparing protein sequences over DNA sequences when these are available, and comparing protein crystal structures. However, there are difficulties with these comparisons as well, not the least of which is the availability of the data on ancestral sequences, protein sequences, or crystalline structures. To further complicate the issue, how does one determine what qualifies as an ancestral sequence?

We see then that ties between different genes or proteins are being constructed on the basis of judgments of what counts and what does not count as "reasonable" or "close enough." While convergence is always a potential explanation for a high similarity score, biologists often make judgments saying that the results are not due to convergence or randomness but to common ancestry. This judgment is often backed by probabilistic arguments.[56] However, probabilistic arguments are made on the basis of assumptions that cannot be checked. Statistical measures of similarity due to nonrandomness are also the product of local decision making. Although there are other kinds of evidence that might help to confirm homology from a high similarity score (however that is decided), such proofs are often not available.

My interest here is to illuminate the processes through which the semantic confusion in evolutionary biology between similarity and homology is being reconstructed in contemporary molecular biological and bioinformatics practices, as they pose and confirm theories of function (see example in the next section). My point is that relatedness among DNA and protein sequences are the products of concepts and tools that have diverse meanings in different situations and that they require judgment calls to be made locally by researchers in each situation. The lack of distinction in the use of the words "homology," "similarity," and "identity" that I have noticed in various laboratories is emblematic of the assumptions being made about homology based on evidence of "strong similarity," which itself is problematic. This equivalence of the terms similarity and homology is not news to molecular biologists. They attempt, when possible, to get other kinds of evidence to confirm homologies. However, they also understand that they are often making assumptions that they cannot confirm with other data.

The important question here is *whether or not it matters* whether one calls a similarity a "homology" or simply a "similarity" in molecular biology and bioinformatics. Is this just semantic confusion or something more consequential for theory construction? Are the local judgments that occur in the construction of homologies simply another example of a phenomenon common to any science in which ambiguities prevail? To explore this point, I present a case where sequence data were used to construct homologies across species, which, in turn, were used to construct a theory.

Constructing Theories of Function from Sequence Similarities in Molecular Biology

Evolutionary biologists and molecular biologists often care about and attend to issues and problems of homology in different ways. An important job in evolutionary biology is the construction of evolutionary trees or phylogenies. Whether similarities are due to convergence or common ancestry is a central question to their construction of genealogies. In contrast, molecular biologists and biochemists tend to be interested more in functional similarities (that is, similarities in the function of a gene sequence and its protein product) than in common ancestry. It is their usual practice to "check" high similarity scores from computer analyses through "wet lab" experimental analyses to confirm or disconfirm functional similarity.[57] They express the gene (have it produce a protein product in the laboratory) and then observe the protein's function in cells. If they find similarities in function at the protein level, they are satisfied.

However, molecular biologists do care about common ancestry in cases when they use homologies, rather than experimentation, to construct and confirm theories about genetic function. In cases where high – but not high enough – similarity scores are not enough to convince other scientists of a common function, and when other kinds of evidence are not (yet) available, molecular biologists have been known to use homology defined as common ancestry as their basis for argument. A good example comes from oncogene research. In the following case, researchers used strong similarities in DNA sequences from different vertebrate species from fish to primates (including humans) to infer gene conservation through evolutionary history. This inference in turn relies on an inference about common ancestry, since the retention of forms implies a direct line of relationship between organisms.

In the late 1970s and early 1980s, by borrowing and using the concept of gene conservation from evolutionary biology, tumor virologists extended their research on viral oncogenes to develop the concept of normal cellular genes as causing human cancers. During the 1960s and early 1970s, tumor virologists reported that they had found specific "cancer" genes in the viruses, which transformed cultured cells and caused tumors in laboratory animals. This experimental work was done using traditional virological and molecular biological methods to investigate RNA tumor viruses[58]. As more researchers became interested in this line of research and explored other viruses, they reported "discoveries" of more viral oncogenes. These viral oncogenes, however, caused cancer only *in vitro* and in laboratory animals. No naturally occurring tumors in animal and human populations were credited to viral oncogenes.

In 1976, J. Michael Bishop, Harold T. Varmus, and their colleagues at the University of California, San Francisco, announced that they had found a *normal* cellular gene sequence in various normal cells of several avian species, a sequence very similar in structure to the chicken viral oncogene, called *src*.[59] Two years later, after using recombinant DNA technologies to construct a probe for their viral oncogene, they also reported that they had discovered DNA sequences related to the *src* viral oncogene in the DNA of normal cells in many different vertebrate species from fish to primates, including humans.[60] Bishop and Varmus suggested that the viral gene causing cancer in animals was transduced from normal cellular genes by the virus; that is, the virus took part of the cellular gene and made it part of its own genetic structure. Based on their and others' research, Bishop and Varmus speculated that some qualitative alteration (through point mutation, amplification, or chromosomal translocation) of this normal cellular gene may play an important role as a cause of human cancer. Before this research, decades of efforts to link viruses to human cancer had been unsuccessful.

In 1978, Bishop and Varmus proposed that the gene that caused normal cells to become cancer cells was part of the cell's normal genetic endowment. They used evolutionary arguments to back up their claim. Since the gene was found in fish, which are evolutionarily quite ancient, the gene must have been conserved through half a billion years of evolution. Their critics simultaneously based their concerns on the theory's "evolutionary illogic." Why would a cancer gene be conserved through evolution? At the time, the announcement of normal cellular genes homologous to a viral oncogene in humans was greeted with some skepticism. According to Bishop,

the first couple of years [after the discovery] were difficult. [Our findings that viral oncogenes had *homologous* sequences in normal cellular genes] were extended with some difficulty to a second and third gene..., and then it was rapidly extended to all the rest [of the 20 known viral oncogenes]. We had to overcome a bias in the field. Our findings were first.... Well, they were rationalized. It was hard for us to come to grips with the idea that a gene carried by a chicken virus that caused cancer was also in human beings. It didn't make sense. *Why would we have cancer genes as part of our evolutionary dowry?* [61]

Bishop saw this resistance as an example of anthropomorphism, which was finally overcome only by experimental evidence that showed the "same" oncogene could be found in humans.

Our first evidence that human beings had this gene, although it evolutionarily looked just fine, there are a lot of biologists, who don't really accept the evolutionary logic.... So until the gene was isolated from humans and shown to be the same as what we'd started with, there was still some doubt. At the outset, there was a lot of skepticism as to whether we had really found the same gene in human beings. That's an anthropomorphism that amused me. Everyone was perfectly happy that the gene was in chickens or even mice, but wasn't supposed to be in humans. I don't know why. But there was a lot of resistance to that.... [62]

Their theory was objected to as evolutionarily illogical because it proposed the conservation of genes that could cause cancer. In order to explain the conservation, as well as the inferred "evolutionary success,"[63] of potential "oncogenic" gene sequences in many different animal species, Bishop and Varmus proposed that their "normal" cellular proto-oncogene had something to do with normal cell division. Later molecular biological and biochemical research on normal growth and development proposed the existence of growth factor genes based on growth factor protein research. Bishop and Varmus began to examine potential relationships between their work on oncogenes and concurrent studies of growth factor proteins. In separate efforts, Doolittle et al. and Waterfield et al. searched sequence databases to find matches between platelet-derived growth factor (PDGF) and a partial sequence of the protein product of the *sis* oncogene of simian sarcoma virus.[64] This computerized search and its subsequent work created a link between normal growth research in developmental biology with cancer studies. In 1984, Waterfield's laboratory reported that they had found that the epidermal growth factor (EFG) receptor protein was "identical" to an oncogene's (*erbB*) protein product studied by the Varmus and Bishop group.

These similarities between normal growth factor gene sequences and proto-oncogene sequences now provided what others considered to be an evolutionarily acceptable (logical) reason for finding that potentially cancer-causing genes were conserved through time: that is, that a "normal" proto-oncogene and its "normal" functional role in growth and development was conserved through time. The oncogenic sequence was a kind of evolutionary relative to the normal growth gene.

The logic of evolution would not permit the survival of solely noxious genes. Powerful selective forces must have been at work to assure the conservation of proto-oncogenes throughout the diversification of metazoan phyla. Yet we know nothing of why these genes have been conserved, only that they are expressed in a variety of tissues and at various points during growth and development, that they are likely to represent a diverse set of biochemical functions, and that they may have all originated from one or a very few founder genes. Perhaps the proteins these genes encode are components of an interdigitating network that controls the growth of individual cells during the course of differentiation. We are badly in need of genetic tools to approach these issues, tools that may be forthcoming from the discovery of proto-oncogenes in Drosophila and nematodes.[65]

In this case, practices and concepts in both molecular biology and evolutionary biology were employed in theorizing about molecular genetic function. The molecular biologists used the concept of "homology," including its common ancestry component (and not just similarity), to construct theories of function about oncogenesis and normal growth and development. Common ancestry was not simply incidental to their claim, since other theories could explain the high similarity scores.

Discussion: Representation, Standardization, and Power

The various definitions of, and debates over, the concept of homology, then, have consequences in the theories, practices, and projects that create our visions of nature. More specifically, these debates are important because they highlight the processes through which the new bioinformatics is creating new images of nature and humanity. These images are in part expressed in what scientists call a new "language" of DNA and amino acid sequences. The "vocabulary" of this "language" is constructed from "homologous" sequences, that is, similar sequences that share common ancestry.

Sequence information and their databases make up a crucial component in the remaking of much of biology in the past decade, and so they will in the years to come. They are being built through the commitment of large resources, and through work of many individuals, organizations, institutions, and tax dollars. They are changing the way biology is practiced, changing the relationships between biologists, computer scientists, clinicians, etc., and changing what can count as durable claims to knowledge about the world.

Information science analyses across laboratories and their experimentally produced representations have provided, according to molecular biologists, unexpected and useful linkages, as the homology between an oncogene and a normal growth factor gene demonstrates.[66] But how was the concept of homology specifically used to construct the theory of oncogenes, and more broadly, how is it being used to construct theories in molecular biology and bioinformatics, when the uses of homology are clearly constructed locally?

In order to answer this question, I will first discuss the issues of representation and standardization in molecular biology and the new bioinformation sciences.

Representation

Representation has been a definitive concept in the social and cultural studies of science.[67] I focus here on ethnographic studies of science, which have documented the local, contingent and situated details of scientific practice.[68] These studies looked at the practices that go into creating "facts" and artifacts (including instruments, materials, and texts). Final products were shown to be selected and shaped through the processes of simplification and deletion of much of the work practice. The contingencies, the mistakes, the blind alleys, the fiddling with and coddling of materials, the magic, black art, and tacit knowledge, the self-styled and adjusted and shortened versions of "cookbook" techniques have been deleted from or are buried in the final representations.

This particular treatment of representations, especially whether such deletions in scientific work could make a difference, has been debated. Lynch, for one, has argued that the social studies of science have raised representation, defined as "methodological horrors," to the status of ideology.[69] Lynch uses Woolgar[70] to describe this treatment of representation.

For Woolgar... the "methodological horrors" are a set of problems raised by sceptical treatments of representation. These include the indeterminate relationships between rule and application, and between theory and experimental data. Woolgar gives a methodological rationale for his global scepticism about scientists' representational practices. The policy of unrestricted scepticism licenses the sociological "observer" to impute methodological horrors to practices that would otherwise appear unperturbed. This interpretive policy requires us to envision a picture of scientists endlessly laboring to evade or circumvent the problems a sceptical philosopher could raise... . Woolgar (p. 101) states that 'science is no more than an especially visible manifestation of the ideology of representation.' The latter he defines (p. 99) as 'the set of beliefs and practices stemming from the notion that objects (meanings, motives, things) underlie or pre-exist the surface signs (documents, appearances) which give rise to them.'[71]

Similarly for Lynch, representations do not "correspond" to objects.

To claim that our investigations reveal deficiencies in representational practice in science would be to assume a correspondence between argument and object as our ideal representational aim.... [T]he very idea of deficiency implies the availability of objects which are somehow free of representation. On the contrary, our position is that representations and objects are inextricably interconnected; that objects can only be 'known' through representation. Criticism necessarily involves competition between representations, not between representation and an 'actual object.'[72]

A scientific representation can only be understood within its context of production and use. In Suchman's words, "we must understand [scientific representations] in relation to, as the product of and resource for, situated practice."[73] In order to truly understand the meaning of a representation, then, we would have to study the contextualized production and use of a representation in each local situation, with its own set of practitioners, skills, assumptions, equipment, social organization, administration, and so on.

How we define representation makes a difference for the second problem that I raise in this paper: that of continuity between knowledge production sites. If we accept that "objects" can only be "known" through representations and that representations and meanings are always constructed in context, what does it mean for biologists in developmental biology to relate their growth hormone gene sequences to oncogene sequences in terms of homologies? Take the example of sequence information representing a biological molecule, which behaves according to a complex set of interactions in its context in a particular location in an organism. What does it mean to represent the molecule with a string of letters that signify, say, a standard understanding of that molecule?[74] According to Lynch, it does not "mean" any more than it is: a string of letters that signify a standard understanding of that molecule. There is no meaning inherent in a signifier. Meaning is only assigned when the signifier is inserted into a particular set of practical activities. In ethnomethodology, this definition is referred to as "the indexicality of language".[75]

What does this understanding of representation mean for the constitution of facts in molecular biology and bioinformatics? Consider that a string of C, A, T, Gs signify a standard understanding that allows scientists to interpret these as cytosine, adenine, thymine, and guanine molecules. These names are understood by most biologists as the molecules that form, in various combinations, genes or sequences that "code for" particular protein products. This, of course, is the point of contention between current, so-called standard biological understandings and the criticisms of biology coming from writers such as Donna Haraway, Ruth Hubbard, Lily Kay, Evelyn Fox Keller, and Richard Lewontin.[76] They criticize this view of DNA as the "master molecule" that determines phenotype and argue for the irreducibility of phenotype to genotype.[77] In contrast, Lynch would argue that "codes" as representations do not have meaning in and of themselves. "Codes" like A, C, G, T do not make any sense without first taking into account the entire system of molecular biological practice. In a similar fashion, rules, for Lynch imply a context of use in which they function as such. The very idea that an expression or line of text is a "rule" implies such a context.

However, techno-scientific representations such as codes do not simply sit in the laboratories. They also move to other parts of the world and change it in very particular ways.[78] For this reason, I am interested in the work that "codes" do in the world. In the case of cognitive and computer sciences, for example, Lucy Suchman argues that cognitive scientific representations or artifacts have political outcomes. That is, cognitive scientists produce computer programs that represent "the embodied, contingent rationality of scientists' situated inquiries" – their day-to-day practices – as rational action, which they are not.

> While the rational artifacts of cognitive scientists' work are programs that run, cognitive scientists' own rationality is an achievement of practices that are only post hoc reducible to either general or specific representation. Canonical descriptions do not capture "the innumerable and singular situations of day to day inquiry"... The consequence is a disparity between the embodied, contingent rationality of scientists' situated inquiries and the abstract, parameterized constructs of rational behavior represented in computer programs understood to be intelligent. To the extent that cognitive science defines the terms of rational action the disparity is not only theoretically interesting but has political implications as well.... In the case of cognitive science...the phenomena are just those things on which our studies take a stand; namely, the organization of practice.... [W]e have a vested interest – not only in the products of cognitive scientists' theorizing but in the adequate rendering of their and others' situated practice.[79]

Standardization

Standardization involves two sets of practices that are often confounded. First, the sciences often invest incredible resources in standardizing technologies and materials. Second, some technologies come to be the standard way (the right tool) to conceive of and to study particular problems.

Here I want to take an example from the history of cancer genetics to describe what I mean by standardization. In the 1930s and 1940s, Clarence Little and his colleagues standardized mouse colonies through inbreeding, and these inbred lines of experimental animals came to be the standard way (the right tool) to conceive of and to study cancer genetics as well as genetics.[80] Inbred experimental mouse lines "realized" the concepts of genetic inheritance and genetic invariability for biologists interested in "controlling for" genetic "factors." By this I mean that the ideas of invariability, homogeneity, and standardization became real laboratory (arti)facts through the development of these inbred animals (and their tumors). These experimental inbred mice (and other animals) and their tumors transformed practices in multiple sites. From this transformation of practices emerged a new technical (versus abstract) definition of

genetics and cancer. The animals and the collective work behind them
defined a new technical work space for "realizing" the concept of
genetics in cancer research. These animals represented and embodied the
commitments of the cancer research community to a particular definition
of genetics. They were artifacts created under artificial conditions
according to a formula that combined a commitment to rational scientific
principles with a particular conception of cancer as a genetically caused
or transmitted disease.

The resulting animals were not simply experimental animals but
instead *standardized experimental systems.* These "standardized" animals
were used to construct representations that were comparable between
laboratories. They were used to reconstruct laboratory work practices
and, in turn, experimentally produced representations. These practices
and representations were assumed to be homogeneous across laboratories
and through time.[81]

Scientific technologies such as these inbred animal experimental
systems or genetic sequence analytic technologies are not neutral objects
through which nature is laid bare. Instead, they are meaning-laden and
meaning-generating tools, just as language and writing are meaning-
laden and meaning-generating media, not simply vehicles for expressing
meaning.[82] *These scientific technologies are highly elaborated techno-
symbolic systems*, just as artificially intelligent computer and robotic
systems are elaborated symbolic systems. That is, scientists "create"
nature in laboratories just as they create "intelligence" in computers.
They create "nature" along the lines of particular commitments and with
particular constraints, just as computer scientists create computer technologies along the lines of particular commitments and with particular
constraints. Thus the boundaries between science, nature, and technology
are illusory. To bring this theoretical discussion back to the case of inbred
mice, geneticists employed experimental technologies and protocols to
create novel scientific/technological objects. These scientific and
technological objects *are* nature, at least as nature is defined by scientific
prescripts. Little's and DeOme's inbred animals cum experimental
systems embody *the phenomena;* that is, they give materiality to objects
such as "genes," "viruses," "cancer," as well as to scientific ideas such as
"genetic invariance," "reproducibility," and "facts".

One could argue that the constructed inbred mouse cum experimental
system limit opportunities to make inferences and extrapolations from
experimental studies to "real natural" situations. Scientists talk about this
problem as one of the limits of generalizability of their findings. In

contrast, theorist and historian of science Donna Haraway argues, in contrast, that there is no real world "nature" that we can know apart from our "technical-natural" tools.[83] These experimental systems *reinvent* nature as they incorporate sociocultural understandings.

Whichever position one takes with respect to their relationship to "nature," the inbred lines of experimental mice were clearly very different from mice outside the laboratory. They were the result of much ingenuity, money, and purposeful as well as serendipitous and improvised efforts by researchers. They were novel phenomena constructed by scientists in their efforts to study a "controlled" "nature" in the laboratory. (Scientists recognize the differences between the "artifactual nature" they create within the laboratory and the "extra-laboratory nature," but argue for their relevance to "actual natural conditions." Other scientists debate and criticize this extrapolation.)[84]

Experimental systems are the result of collective and historically situated processes involving and implicating a broad set of actors, time, and spaces. I assume that scientific theories, facts, and technologies are produced through the collective (sometimes cooperating, sometimes conflicting, sometimes indifferent) efforts and actions of many different scientists, groups, tools, laboratories, institutions, funding agencies, janitors, companies, social movements, international competitions, etc. Instead of speaking of work within or outside an experimental framework, of forces acting from outside to affect or create changes in something inside an experimental framework, I speak of an experimental system as incorporating and representing particular commitments.

To summarize, I define standardization in two ways. First, after technologies, concepts, and laboratories have become standardized, the standards rhetorically and literally allow laboratories across the biological laboratories to compare and contrast outcomes produced in different sites. Second, after they become installed as the "standard" tool, they shape future collective commitments and actions. Thus technologies and materials are wed to theoretical problems and hypotheses, but in ways that are more complex than some scientists acknowledge. Scientific technologies are highly elaborated symbolic systems, not neutral media for "knowing" nature. For example, neutrality, or the idea that one can eliminate "noise" versus "signal" to reach a *tabula rasa* from which one can then produce "reproducible effects," is part of a set of "values" historically located in so-called "Western traditions of thought." These values include realist, objectivist, and empiricist rhetorics, which form the basis for establishing factness and the universality of findings.

Power: Universalizing via Standardization

A question posed in this paper is whether or not some technologies and practices of representation, some representational devices, are constructed as more authoritative than others.[85] For example, do "metrologies" or standardized instruments, techniques, and codes[86] contribute to acceptance and stabilization? In the case of molecular genetic sequence information, does the selection of four standard molecules in a linear readout of A, C, T, Gs for DNA and 20 standard molecules in the linear readout of A, R, N, etc., for the single letter codes for amino acids provide a better "substrate" for constructing authoritative representations than the complexities introduced by critics, including Hubbard, Lewontin, et al.?

Through fierce struggles, molecular genetic sequence information has become the authoritative way of representing genes and bodies.[87] Fujimura and Chou discuss the practices that have, through many struggles, come to be the authoritative style of doing science.[88] Hacking discusses the reasoning styles that have developed through time to become the authoritative rules for authenticating theories and facts.[89]

The debates about homology tell us that "homology" is a set of competing technologies for creating representations of relatedness between different natural objects. These debates are part of an ongoing struggle. According to Donoghue, "the history of the word homology can be interpreted as a series of responses to challenges brought on by underlying conceptual changes."[90] He then proceeds to give some examples of different interpretations and ends with this conclusion: "The choice of a definition is, at least in part, a means of forcing other scientists to pay closer attention to whatever one thinks is the most important.... For better or for worse, the choice of definitions helps determine whose agenda will attract the most attention."[91]

If we assume that molecular genetic sequences are very particular and singular kinds of representations of nature, then any representation created via the comparisons of these standardized entities using the concept of homology should also be understood to be only one way of "knowing" an object. I am interested in the competition between representations and how one representation becomes more powerful than another. I argue that at least four steps taken in the language of standardization are being used by molecular geneticists to promote their theories as powerful tools for representing nature. The first step is the representation of genes as strings of A, T, C, Gs. The second step is the effort to produce a "consensus homology," an algorithm that will be used by everyone to determine when similarity achieves the status of homology.[92] (This is an ongoing set of negotiations, confusions, and struggles, as we have seen.) The third step is to upload these new "reagents" in computerized databases and these homology algorithms (and other algorithms) in computerized software. The final step is the promotion of these steps as meeting, and contributing to, the realist, objectivist, and empiricist rhetorics of knowledge construction in contemporary science.

Molecular genetic sequences as reagents provide one form of standardization that, rhetorically and literally, allow laboratories across the biological subdisciplines to compare and contrast outcomes produced in these different sites. This form of standardization is used by scientists to argue that their technologies are creating more objective knowledge because they are comparing "like" entities when using concepts such as homology to construct relations between these "standard reagents." That is, if biologists have standardized their reagents to the "same" molecules, then the differences or similarities between them can be treated as significant findings. That is, this form of standardization means that any similarities or differences that are found are attributed by scientists to "real" similarities or differences rather than to "noise" or random occurrences.

Conclusion: Power and the Production of the Genetically Normal

We are living through a movement from an organic, industrial society to a polymorphous, information system – from all work to all play, a deadly game. Simultaneously material and ideological, the dichotomies may be expressed [as a set] of transitions from the comfortable old hierarchical dominations to the scary new networks I have called the informatics of domination.[93]

Disease is a subspecies of information malfunction or communications pathology.[94]

The rhetoric of genome researchers is often framed in the terms of disease diagnosis and treatment. They argue that the resources invested and the potential concerns (ethical, legal, and social) are a small price to pay for the future possibility of recreating bodies in "healthier" forms. The question raised by many social scientists has been "Who gets to determine what is 'healthier?'" Who decides, and by which criteria, that something is a genetic disease, disorder, or defect?

One of my students discussed her inverted chromosome with me. Her mother was pregnant with her fifth child when she was past the age of 35. The doctor recommended amniocentesis, and the results showed that the fetus had an inverted chromosome. The doctor recommended abortion. Her mother decided not to abort on the basis of a guess that perhaps all her children had an inverted chromosome. (Her mother is a chemist.) The four older children, include "Jessie," were then tested and several more found to have an inverted chromosome. "Jessie" then suggested that had that test been available when she, the eldest child, was in utero, and had her mother yielded to the doctor's advice, she might not have been born.

There is much written by sociologists about specific concerns surrounding genetic screening. I want to focus my discussion more broadly on the consequences of genetic diagnostic, screening, and intervention technologies. I want to raise the general issue of the construction of normality, disease/defect, and the historical subject. Michel Foucault redefined the notions of power, subject, and history. He examined the strategies and technologies through which people become disciplined, through which they become docile. For Foucault, power becomes located in these strategies and technologies. It is in these technologies that power is masked well enough for it not to be "cynical" (overt), and therefore for it to be successful. Rather than look to law as a model or code, he constructed his analytics of power through the discourses and technologies of sexuality, of the clinic, of the prison, of classifications and taxonomies of scientific knowledge.[95]

For Foucault, the interesting power is not that which is located in a group of institutions and mechanisms, in explicit rules and codes, or in "a general system of domination exerted by one group over another." "Power is not an institution, and not a structure; neither is it a certain strength we are endowed with. It is the name that one attributes to a complex strategical situation in a particular society."[96] Disciplinary power is his object of study. It is only after behavior has already been "disciplined" and "normalized" via these strategies that laws are then used to regulate people and their actions.

Foucault is concerned with the construction of this power, in the forms of the "sovereign," the "normal," and the "objective." This happens, in most cases, not through the explicit use of force or the explicit adjudication of law, but through the historical investing of particular orders of power.

It is a question of orienting ourselves to a conception of power which replaces the privilege of the law with the viewpoint of the objective, the privilege of prohibition with the viewpoint of tactical efficacy, the privilege of sovereignty with the analysis of a multiple and mobile field of force relations wherein far-reaching, but never completely stable, effects of domination are produced.[97]

In the example of genomics and the development and proliferation of new genetic technologies into medical practices, I want to put at the forefront of our discussions a concern with the production of the normal and the pathological. Oftentimes, sociologists and anthropologists writing about the new genetic technologies discuss power in terms of existing social and economic hierarchies, organizations, and institutions. But these existing hierarchies should be examined with relation to the constitution of the normal and pathological via technologies, practices, discourses. Social and economic hierarchies (hegemonies), organizations, and institutions police or regulate deviations from the normal. However, the more powerful regulation, in the form of what is acceptable, has already been put into place in our very understandings of "health" and "illness," the "normal" and the "pathological."

In many ways, Foucault's normalization parallels the concept of standardization that I use here. In nineteenth and twentieth century scientific research, standardized experimental analytic systems have created the "nature" and the "society" we "know". Normalization is accomplished through historical processes, which are not to be taken for granted as the only ways in which nature and society could be produced. However, once nature and society are produced in particular forms, they are often assumed to be the only forms in which nature and society could exist. That is, these historically and socially produced forms become "naturalized."

A consequence of naturalization and normalization strategies is the creation of new kinds of bodies and new kinds of subjects. Foucault's subject is not the individual of methodological individualism. Instead, his subject is constituted by a particular combination of discourses located in time and place. The self-consciousness, motivations, desires, and actions of this new subject are constituted from these strategies. Similarly, the body created by the naturalizing and normalizing strategies of new genetic technologies are "material-semiotic nodes."

Bodies...are not born; they are made. Bodies have been as thoroughly denaturalized as sign, context, and time.... Always radically historically specific, bodies have a different kind of specificity and effectivity, and so they invite a different kind of engagement and intervention. The notion of a 'material-semiotic actor' is intended to highlight the object of knowledge as an active part of the apparatus of bodily production, without *ever* implying immediate presence of such objects or...their final or unique determination of what can count as objective knowledge of a biomedical body at a particular historical junction. Bodies as objects are material-semiotic generative nodes. Their boundaries materialize in social interaction;

'objects' like bodies do not pre-exist as much. Scientific objectivity...is not about dis-engaged discovery, but about mutual and usually unequal structure.... The various contending biological bodies emerge at the intersection of biological research, writing, and publishing; medical and other business practices; cultural productions of all kinds, including available metaphors and narratives; and technology...[98]

Now consider the transformation of women diagnosed with the newly located breast cancer gene, BRCA1, into bodies with pathological genes. A recent newspaper article quoted the researcher whose laboratory had "located" the BRCA1. He referred to these women as "carriers." I argue that the term "carrier" is part of the language of molecular genetic informatics strategies. This language symbolizes both a conception of these women as genetically pathological and a commitment to particular paths of action. "Language is not about description, but about commitment."[99] Our choice of languages, then, is a choice about the commitments we want to make and the futures we want to build.

Acknowledgement

An early version of this paper was presented at the Sommer Akademie Conference on Communicating Nature, Berlin, Germany, July 23-29, 1994. It has benefited from the discussion at the Sommer Akademie and from comments by Sören Germer, Mimi Ito, Alberto Cambrosio, Carolyn Laub, Michael Lynch, Ray McDermott, Susan Newman, and Lucy Suchman.

Notes

1. Gunnar von Heijne, *Sequence Analysis in Molecular Biology: Treasure Trove or Trivial Pursuit.* (San Diego: Academic Press Inc., 1987), p. 151.
2. Harold J. Morowitz and Temple Smith, *Report of the Matrix of Biological Knowledge Workshop.* (Santa Fe, NM: Santa Fe Institute, July 13-August 14, 1987), pp. 1–2.
3. Excerpt from a letter of the president of Johns Hopkins, December 1991.
4. David B, Wake, "Comparative Terminology," *Science, 265,* July 8 1994, 268.
5. Robert Pollack, *Signs of Life: The Language and Meanings of DNA* (Boston: Houghton Mifflin Company,1994), p. 159.
6. For histories of the Human Genome Initiative see Charles R. Cantor, "Orchestrating the Human Genome Project," *Science, 248,* April 6 1990, pp 49–51; Robert M. Cook-Deegan, *The Gene Wars: Science, Politics, and the Human Genome,* (New York: W.W. Norton & Co., 1994); Michael Fortun, "Mapping and Making Genes and Histories: The Genomics Project in the United States, 1980-1990," doctoral, History of Science Department, Harvard University, 1993; and Judson (1992); James D.

Watson, "The Human Genome Project: Past, Present, and Future," *Science, 249,* April 6 1990, pp. 44 – 49.

7. See quotes from other biologists at the beginning of this paper. There is an "old" theoretical biology, which has existed for some time, but which has not had the reputation of, say, theoretical physics. See von Heijne, *Sequence Analysis in Molecular Biology.* As historians of science have noted, this is not a novel move. The transformation of the body into an information system began in the 1940s, and has continued since that time, with the development of computers, cybernetics research, information theory, and military maneuvers to build command-and-control systems. The delineation of the structure of DNA was another step in this process of the informatizing of the body, but not the beginning of the process. See Donna J. Haraway, "The High Cost of Information in Post-World War II Evolutionary Biology: Ergonomics, Semiotics, and the Sociobiology of Communication Systems," *The Philosophical Forum, XIII,* Winter-Spring 1981-1982, pp. 244 – 278; Evelyn Fox Keller, "The Body of a New Machine: Situating the Organism Between Telegraphs and Computers," in *Refiguring Life: Metaphors of Twentieth Century Biology* (New York: Columbia University Press, 1995); Lily Kay, "Who Wrote the Book of Life? Information and the Transformation of Molecular Biology, 1945 – 55," in *Experimentalsysteme in den Biologische-Medizinische Wissenschaften: Objekt, Differenzen, Konjunkturen,* ed. Michael Hagner and Hans-Jorg Rheinberger (Berlin: Akademie Verlag, 1994); and Rich Doyle, "Mr. Schrodinger Inside Himself? The Rhetorical Origins of the Genetic Code," Chapter 2 (Department of Rhetoric, University of California, Berkeley, doctoral dissertation, 1993).

8. Walter Gilbert, "Towards a Paradigm Shift in Biology," *Nature, 349,* Jan 10 1991, 99.

9. Gilbert also argued that this paradigm shift from experimental biology to an information-based theoretical science will cure the "malaise" of technologization that has befallen biology. Many molecular biologists, as well as many graduate students in molecular biology, agree with Gilbert that molecular biology has become a series of cookbook techniques, complete with recipes and kits. Indeed, Gilbert's own development of fast and relatively simple sequencing techniques contributed to the technologization of molecular biology. See Joan H. Fujimura, "The Molecular Biological Bandwagon in Cancer Research: Where Social Worlds Meet," *Social Problems, 35,* (1988), pp. 261–283.

10. Gilbert, "Towards a Paradigm Shift in Biology," p. 99.

11. Von Heijne, *Sequence Analysis in Molecular Biology,* p. 151.

12. See Walter Gilbert, "Genes-in-Pieces Revisited," *Science, 228* (1985), pp. 823–824; J. Rogers, "Exon Shuffling and Intron Insertion in Serine Protease Genes," *Nature, 315* (1985), pp. 458–459; and W. Gilbert, M. Marchionni, and G. McKnight, "On the Antiquity of Introns," *Cell, 46* (1986), pp. 151–154.

13. Von Heijne, *Sequence Analysis in Molecular Biology,* p 40. For another example, see T.F. Smith, A. Srinivasan, G. Schochetman, M. Marcus, and G. Myers, "The Phylogenetic History of Immunodeficiency Viruses," *Nature, 333,* June 1988, pp. 573–575. Hillis refers to exon shuffling as "partial homology." David M. Hillis, "Homology in Molecular Biology," in *Homology: The Hierarchical Basis of Comparative Biology* ed. Brian K. Hall. (New York: Academic Press, 1944), pp. 339 – 368.

14. CMSHG *Report of the Committee on Mapping and Sequencing the Human Genome, Board on Basic Biology,* Commission on Life Sciences, National Research Council, 1988, p. 55.

15. Robert L. Dorit, Lloyd Schoenbach, and Walter Gilbert, "How Big Is the Universe of Exons?" *Science, 250,* Dec 7 1990, pp. 1377–1382. Critiques of the research of Dorit and his colleagues by molecular biologists ranged from arguments with the thesis of the paper to arguments about the detailed assumptions made in constructing the computer program that produced the final numbers.

16. Anthropologists would refer to this as an origin story.

17. More recently, several researchers have argued that so-called "junk DNA" plays critical roles in the organism's activities. See Roy J. Britten, David B. Stout, and Eric H. Davidson, "The Current Source of Human Alu Retroposons Is a Conserved Gene Shared with Old World Monkey," *Proceedings of the National Academy of Sciences, USA, 86,* May 1989, pp. 3718–3722; and Ben F. Koop and Leroy Hood, "Striking Sequence Similarity over Almost 100 Kilobases of Human and Mouse T-cell Receptor DNA," *Nature Genetics, 7,* May 1994, pp. 48 – 53.

18. National Center for Human Genome Research. *Annual Report 1 – FY 1990,* Department of Health and Human Services, Public Health Service, National Institutes of Health, 1990, pp. 30 – 31.

19. I am studying this process in my larger ethnographic research project.

20. NCHGR, *Annual Report 1 – FY 1990,* p. 26. See also Michael S. Waterman, "Genomic Sequence Databases," *Genomics,* 6 (1990), pp. 700 –701.

21. See also Michael Fortun, "Time, Busy Bodies, and the Habit of Becoming Genetic" (forthcoming) on "speeding" as a more general phenomenon in the twentieth century.

22. Sequence databases are only one kind of database. Genetic map databases provide different kinds of information. I limit my discussion for the rest of this paper to sequence databases. Major efforts have also been directed at linking the various databases to provide means for linking protein, DNA, RNA, and genetic map information for each sequence of interest. There is great debate about how to construct the semantics of each database to provide the "best" use of the diverse databases.

23. Not all laboratories have the funds for computer access, and if they do have access, their universities might not have funds available for the requisite computer infrastructure or personnel with the skills to deal with the intricacies of the databases.

24. All of these options require further work, access to new resources, and, therefore, work reorganization. For example, a graduate student sequenced a gene (RNA) of a protist and discovered to his chagrin that he then had to arrange to gain access to the single existing database of RNA sequences for this protist. The database was private – that is, it was developed by and is maintained by an academic researcher famous for his work on the protist. In order to gain access, the graduate student arranged to work as a postdoctoral fellow in the laboratory with his required database. He was not pleased about the arrangement, however. He felt that he had now become "a twig on the [scientist's] tree" in order to be able to make some meaning of his sequence.

25. These figures are for July 1988. See Rita R. Colwell, ed., *Biomolecular Data: A Resource in Transition,* (Oxford: Oxford University Press, 1989), p. 124.

26. *C. elegans* was promoted by Schatz and Roberts as a model for the sequencing and database building of information on the human genome. Bruce R. Schatz, "Building an Electronic Scientific Community," in *Proceedings of the 24th Annual Hawaii International Conference on Systems Science,* III., ed. Jay Nunmaker (Los

Almaritos, CA: IEEE Society Press, 1991); Leslie Roberts, "The Worm Project," *Science, 248* June 15 1990, pp. 1310–1313.

27. Compact Disc Read Only Memory, or CD-ROM, contain digitally encoded information readable by computer. These discs are "read only," that is, they cannot be modified by the user. However, "read and write" CDs will soon be commercially available. It is unclear what impact "read and write" CDs will have on the collective production and use of genome information. An addendum: Walter Gilbert has predicted that, as various human genome projects proceed, each individual will soon be able to have her or his own genome on a CD.

28. Von Heijne, *Sequence Analysis in Molecular Biology*, p. 123.

29. Russell F. Doolittle, *Of Urfs and Orfs: A Primer on How to Analyze Drived Amino Acid Sequences.* (Mill Valley, CA: University Science Books, 1987), p. 25. For an early example, see below.

30. *Ibid.*, p. 17.

31. David B. Wake, "Comparative Terminology," p. 268.

32. *Ibid.*

33. Doolittle, *Of Urfs and Orfs*, pp. 35 – 36.

34. Von Heijne, *Sequence Analysis in Molecular Biology*, p. 123.

35. Hillis, "Homology in Molecular Biology," pp. 340 – 341.

36. Colin Patterson, "Introduction," in *Molecules and Morphology in Evolution: Conflict or Compromise?* ed. C. Patterson (Cambridge: Cambridge University Press, 1987), p. 18.

37. Michael J. Donoghue, "Homology" in *Keywords in Evolutionary Biology,* ed. Evelyn Fox Keller and Elizabeth A. Lloyd (Cambridge, MA: Harvard University Press, 1992), p. 170.

38. Hillis, "Homology in Molecular Biology," p. 340.

39. Donoghue, "Homology," p. 171.

40. *Ibid.* With respect to the second point, differences in form between two animals or species were interpreted to mean that change or transformation had occurred, rather than that the two were unrelated.

41. I cannot do justice to the many complex debates about homology in this short paper. For a more extensive and nuanced view of these complexities, see Brian K. Hall, ed., *Homology: The Hierarchical Basis of Comparative Biology* (New York: Academic Press, 1944).

42. For biologists who criticize this view, see, example. V. Louise Roth, "Within and Between Organisms: Replicators, Lineages, and Homologues," in *Homology: The Hierarchical Basis of Comparative Biology,* ed., Brian K. Hall (New York: Academic Press, 1944), especially pp. 311– 312; Ruth Hubbard, *The Politics of Women's Biology* (New Brunswick: Rutgers University Press, 1990); R. Hubbard and E. Wald. *Exploding the Gene Myth* (Boston: Beacon Press, 1993); and Richard C. Lewontin, *Biology as Ideology: The Doctrine of DNA* (New York: Harper Collins, 1991). For historians of science who criticize this view, see example, Donna J. Haraway, *Simians, Cyborgs and Women: The Reinvention of Nature* (New York: Routledge, 1991); Lily Kay, "Constructing Histories of Twentieth-Century Experimental Life Science: The Promise and Perils of Archives," *The Mendel Newsletter: Archival Resources for the History of Genetics and Allied Sciences,* 2, November 1992; Evelyn Fox Keller, *Reflections on Gender and Science* (New Haven: Yale University Press, 1985); and Evelyn Fox Keller, *Secrets of Life, Secrets of Death: Essays on Language, Gender and Science* (New York: Routledge, 1992).

43. George V. Lauder, "Homology, Form, and Function," in *Homology: The Hierarchical Basis of Comparative Biology*, ed. Brian K. Hall, (New York: Academic Press, 1944), pp. 162–163.
44. Neil H. Shubin "History, Ontogeny, and Evolution of the Archetype," in *Homology: The Hierarchical Basis of Comparative Biology*, ed. Brian K. Hall, (New York: Academic Press, 1944), p. 252.
45. Roth, "Within and Between Organisms," p. 307.
46. Roth quoted in Lauder, "Homology, Form, and Function," p. 163.
47. Lauder, "Homology, Form, and Function," p. 163.
48. *Ibid.*, p. 163.
49. Donoghue, "Homology," p. 173.
50. If a protein sequence is not available, biologists translate the DNA sequences into sequences of amino acids before beginning a homology search. Since there are only four possible nucleotide bases to permute in DNA and RNA sequences, the probabilities of random matches are very high (25% on the average). The 20 amino acids that constitute proteins reduce the probabilities of random matches. For a discussion of other difficulties of determining homologies using DNA sequences, see Russell F. Doolittle, "Similar Amino Acid Sequences: Chance or Common Ancestry?" *Science, 214*, October 9 1981, p. 153.
51. Dorit, Schoenbach, and Gilbert, "How Big Is the Universe of Exons?" pp. 1377–1382.
52. Donoghue, "Homology," p. 173.
53. Gerald R. Reeck, Christoph de Haen, David C. Teller, Russell F. Doolittle, Walter M. Fitch, Richard E. Dickerson, Pierre Chambon, Andrew D. McLachlan, Emanuel Margoliash, Thomas H. Jukes, and Emile Zuckerandl, "Homology" in Proteins and Nucleic Acids: A Terminology Muddle and a Way Out of It" (Letter to the Editor), *Cell, 50*, August 28, 1987, p. 667 (emphasis added).
54. Von Heijne, *Sequence Analysis in Molecular Biology*, p. 137.
55. Doolittle, "Similar Amino Acid Sequences," p. 153.
56. S. Karlin and C. Matessi, "Kin Selection and Altruism," *Proceedings of the Royal Society of London B, 219* (1983), pp. 327–353.
57. I do not mean to elevate "wet lab" work in molecular biology above any other. I present it instead as an alternative.
58. RNA tumor viruses are retroviruses, which have genes constituted of RNA sequences rather than DNA. They replicate by producing a strand of DNA sequences through the activities of an enzyme called reverse transcriptase. Retroviruses have come to be common knowledge through the HIV or human immunodefiency virus that is believed to cause AIDS.
59. D. Stehelin, H.E. Varmus, J.M. Bishop, and P.K. Vogt, "DNA Related to the Transforming Gene(s) of Avian Sarcoma Viruses Is Present in Normal Avian DNA." *Nature, 260* (1976), pp. 170–173.
60. D.H. Spector, H.E. Varmus, and J.M. Bishop, "Nucleotide Sequences Related to the Transforming Gene of Avian Sarcoma Virus Are Present in the DNA of Uninfected Vertebrates," *Proceedings of the National Academy of Sciences U.S.A., 75* (1978), pp. 5023–5027.
61. Bishop interview (emphasis added).
62. Bishop interview.
63. I use quotations around the term "evolutionary success" to indicate the complexity of this term. Evoluationary biology is embroiled in many debates about the units, levels,

and processes of selection. For example, does natural selection operate at the level of the gene, the entire genome, the individual organism, the "group," or the population? Does it operate on DNA sequences, genes, or the structural proteins? What is selection? Are all selection pressures identical in their impact? What about engineering? "Evolutionary success," then, is a highly contextualized set of discussions and debates rather than an explanatory mechanism. See, for example, Elisabeth A. Lloyd, "Unit of Selection," in *Keywords in Evolutionary Biology*, ed. Evelyn Fox Keller and Elisabeth A. Lloyd (Cambridge, MA: Harvard University Press, 1992); Elisabeth A. Lloyd, *The Structure and Confirmation of Evolutionary Theory* (Westport: Greenwood Press, 1988); and Robert N. Brandon, *Adaptation and Environment* (Princeton, NJ: Princeton University Press, 1990) for an overview analysis of "the units of selection" debates.

64. R.F Doolittle, M.W. Hunkapiller, L.E. Hood, S.G. DeVare, K.C. Robbins, et al., "Simian Sacroma Virus *onc* Gene, v-*sis*, Is Derived from the Gene (or Genes) Encoding a Platelet-Derived Growth factor," *Science, 221* (1983), pp. 275 –76. M.D. Waterfield, G.T. Scrace, N. Whittle, P. Stroobant, A. Johnson, A. Wasteson, B. Westermark, J. Huang, T.F. Deuel, "Platelet-Derived Growth Factor is Structurally Related to the Putative Transforming Protein p28[sis] of Simian Sarcoma Virus," *Nature, 304* (1983), pp 33 – 39.

65. J. Michael Bishop, "Cellular Oncogenes and Retroviruses," *Annual Review of Biochemistry, 52* (1983), pp. 347–348.

66. This reminds me of Donna Haraway's argument that new and unexpected kinds of solidarities are being constructed in this age of the domination of informatics; see "A manifesto for Cyborgs: Science, Technology, and Socialist Feminism in the 1980s," in Haraway, *Simians, Cyborgs, and Women*, pp. 149 –181.

67. For a good overview, see Michael Lynch and Steve Woolgar, "Introduction: Sociological Orientations to Representational Practice in Science" in *Representation in Scientific Practice,* ed. Lynch and Woolgar (Cambridge: The MIT Press, 1990), pp. 1–18.

68. See Harry M. Collins, *Changing Order: Replication and Induction in Scientific Practice* (Beverly Hills, CA: Sage, 1985); J. H. Fujimura, "Constructing Doable Problems in Cancer Research: Articulating Alignment," *Social Studies of Science, 17,* May 1987, pp. 257–293; Karin Knorr-Cetina, *The Manufacture of Knowledge* (Oxford: Pergamon Press, 1981); Bruno Latour and Steve Woolgar, *Laboratory Life*: *The Social Construction of Scientific Facts* (Beverly Hills: Sage, 1986 [1979]); Michael Lynch, *Art and Artefact in Laboratory Science* (London: Routledge and Kegan Paul, 1985); Susan Leigh Star, "Simplification in Scientific Work: An Example from Neuroscience Research," *Social Studies of Science, 13* (1983), pp. 205–28; and L. Suchman, *Plans and Situated Action: The Problem of Human-Machine Communication* (Cambridge: Cambridge University Press, 1987).

69. See Michael Lynch, "Representation Is Overrated," *Configurations, 2,* Winter 1994, pp. 137–149; Michael Lynch, *Scientific Practice and Ordinary Action*: *Ethnomethodology and Social Studies of Science* (New York: Cambridge University Press, 1993); and Lynch and Woolgar, "Introduction: Sociological Orientations."

70. Steve Woolgar, "Time and Documents in Researcher Interaction: Some Ways of Making Out What is Happening in Experimental Science," in ed. Michael Lynch and Steve Woolgar, *Representation in Scientific Practice* (Cambridge, MA: The MIT Press, 1990). Note that Lynch criticizes Woolgar in this quote and collaborates with him in the publication of the volume on representation. The "methodological horrors" issue is a point of tension in much SSK work and fuels collective discussions.

71. Lynch, *Scientific Practice and Ordinary Action*, p. 194.

72. Lynch and Woolgar, "Introduction: Sociological Orientations, p. 13.

73. Lucy A. Suchman, "Representing Practice in Cognitive Science," in *Representation in Scientific Practice,* ed. Michael Lynch and Steve Woolgar, p. 318.

74. I use this definition of a string of letters signifying "a standard understanding of that molecule" as my example because, as Lynch himself argues, while "the problem of correlating 'text' and 'context' may have no empirical or methodological solution[,] this does not mean that whenever we talk we perform an unrecognizable activity, or that language is always and inherently ambiguous. Nor does it mean that we can 'construct' the world in which we live any way we like" (Lynch, "Representation Is Overrated," p. 146). Instead, Lynch would believe that molecular biologists would generally agree that a string of G, A, T, C represent molecules of deoxyribonucleic acids, and that a string of A, L, T, N, etc., represent molecules called amino acids.

75. In another line of science studies writings, Collins has similarly argued that the reproduction of "the same" scientific activities is never replicative in the manner that scientific textbooks define replication. This does not mean that there is no similarity. Instead, Collins and his colleagues problematize the construction of similarity (Collins, *Changing Order*).

76. See footnote 42 above.

77. Biologist Richard Lewontin states this argument more baldly: First, DNA is not self-reproducing. Second, it makes nothing. And third, organisms are not determined by it. See "The Dream of the Human Genome" in Lewontin, *Biology as Ideology*.

78. See Donna J. Haraway, "Universal Donors in a Vampire Culture: It's All in the Family. Biological Kinship Categories in the Twentieth-Century United States," in *Reinventing Nature*, ed. William Cronon, (New York: Norton, 1995); Bruno Latour, "Give Me a Laboratory and I Will Raise the World," in *Science Observed*, ed. Karin Knorr-Cetina and Michael Mulkay (Beverly Hills: Sage, 1983); Bruno Latour, *The Pasteurization of France* (Cambridge: Harvard University Press, 1988), and Michel Callon, "Some Elements of a Sociology of Translation: Domestication of the Scallops and the Fishermen of St. Brieuc Bay" in *Power, Action and Belief, Sociological Review Monograph,* ed. John Law (Boston: Routledge and Kegan Paul, 1986).

79. Suchman, "Representing Practice in Cognitive Science," p. 318. See also Suchman, "Do Categories Have Politics: The Language/Action Perspective Reconsidered," *Computer Supported Cooperative Work*, 2 (1994), pp. 177–190.

80. One of the first institutes to develop research colonies for the study of mammalian genetics was the Jackson Laboratory in Bar Harbor, Maine, founded by Clarence C. Little in 1929. See also Joan H. Fujimura, "Tools of the Trade: A Brief History of Standardized Experimental Systems in Classical Genetic and Virological Cancer Research, ca. 1920–1978," *History and Philosophy of the Life Sciences*, 18 (Sept 1996): pp. 3–54.

81. At the turn of the twentieth century, biology was transformed into an experimental and analytic science. Growing commitments across scientific and medical disciplines to the ideals of positivist empiricism translated into ideals of quantifiable and reproducible experimentation in biology. Although naturalist and ecological inquiries continued in a more taxonomic vein, many new fields of biology began to pursue analytic methods of inquiry. Experimentation and analysis became the hallmarks of "science."

82. For work on the relationship between language and molecular genetics, see "The Biopolitics of Postmodern Bodies: Constitutions of Self in Immune Systems

Discourse," in Haraway, *Simians, Cyborgs and Women*; Keller, *Secrets of Life, Secrets of Death*.

83. Haraway, "Universal Donors."
84. See chapter 8 of Joan H. Fujimura, *Crafting Science and Transforming Biology: The Case of Oncogene Research* (Cambridge: Harvard University Press, 1996).
85. "Authority" is itself an historical and contingent outcome. I do not propose an ahistorical, theoretical explanation for the "best" tools for producing "factness."
86. For a molecular biologist's view of language in molecular genetics, see Robert Pollack, *Signs of Life: The Language and Meanings of DNA* (Boston: Houghton Mifflin Company, 1994).
87. Evelyn Fox Keller, *Refiguring Life*.
88. Joan H. Fujimura and Danny Chou, "Dissent in Science: Styles of Scientific Practice and the Controversy Over the Cause of AIDS," *Social Science and Medicine, 38* April 1994, pp. 1017–1036.
89. According to Hacking, each style of reasoning is historically created. "Every style comes into being by little microsocial interactions and negotiations. It is a contingent matter.... Each style has become *what we think of* as a rather timeless canon of objectivity, a standard or model of what it is to be reasonable about this or that type of subject matter. We do not check to see whether mathematical proof or laboratory investigation or statistical 'studies' are the right way to reason: they have become (after fierce struggles) what it is to reason rightly, to be reasonable in this or that domain." Ian Hacking, "The Self-Vindication of the Laboratory Sciences," in *Science as Practice and Culture*, ed. Andrew Pickering. (Chicago: University of Chicago Press, 1992), p. 10 (emphasis added).
90. Donoghue, "Homology," p. 178.
91. *Ibid*. p. 179.
92. This is just a little play on the human genome project's production of a "consensus genome."
93. Donna J. Haraway, "A Manifesto for Cyborgs," p. 161.
94. Haraway, "The Biopolitics of Postmodern Bodies," p. 212.
95. Michel Foucault, *The History of Sexuality, 1* (New York: Pantheon, 1978); *Discipline and Punish* (New York: Vintage Press, 1975); *The Order of Things: An Archaeology of the Human Sciences*, a translation of *Les Mots et Les Choses* (New York: Random House, 1970); and *The Birth of the Clinic: An Archaeology of Medical Perception* (New York: Vintage Books, 1975[1963]).
96. Foucault, *History of Sexuality*, p. 93.
97. *Ibid.*, p. 102.
98. Haraway, "The Biopolitics of Postmodern Bodies," p. 208.
99. *Ibid.*, p. 214.

CIRCULATING MICE AND VIRUSES

The Jackson Memorial Laboratory, the National Cancer Institute,

and the Genetics of Breast Cancer, 1930–1965

JEAN-PAUL GAUDILLIERE

Hôpital Necker-Enfants Malades, Paris, France

Hegel remarks somewhere that all facts and personages of great importance in world history occur, as it were, twice. He forgot to add: the first time as a tragedy, the second as farce. Cousidières for Danton, Louis Blanc for Robespierre, the *Montagne* of 1848 to 1851 for the *Montagne* of 1793 to 1795, the Nephew for the Uncle.
Karl Marx. The Eighteenth Brumaire of Louis Bonaparte

Mass production is not merely quantity production, for this may have none of the requisites of mass production. Nor it is merely machine production, which also may exist without any resemblance to mass production. Mass production is the focusing upon a manufacturing project of the principles of power, accuracy, economy, system, continuity, and speed.
Henry Ford. "Mass production," Encyclopedia Britannica.

Introduction

The issue of *Nature Genetics* published in October 1992 included a review on breast cancer genes. There M.C. King argued that breast cancer is one of the most common genetic diseases in the industralized world[1]. Such an interest in genetic predispositions to the formation of mammary tumor was hardly visible in the 1960s and 1970s. Debates on the genetics of cancer were more fashionable before World War II, however. For example, in 1932, M.T. Macklin, a preeminent advocate of eugenics, was asking physicians to pay serious attention to "evidence that tumors, both benign and malignant, are inherited in man".[2]

Pointing out similarities between two articles 50 years apart is dubious historical practice since cultural contexts, and scientific and medical practices have changed radically in between. A quick survey of the

Michael Fortun and Everett Mendelsohn (eds.), The Practices of Human Genetics, 89–124
©1999 *Kluwer Academic Publishers. Printed in Great Britain.*

twentieth century literature dealing with experimental studies of cancer suggests nonetheless that the existence of two waves of enthusiasm for the application of genetics to cancer studies and cancer control is not pure imagination. From the 1930s to the late 1950s, *Index Medicus* had an entry for "cancer heredity". In the 1960s, the classification scheme changed: cancer heredity was dropped, and an entry for "tumor viruses" was established. In the 1970s, there was a growing tendency to divide up the "cancer" division into different categories: "breast cancer," "bone cancer," etc. Each category was further divided into specific topics. One of them was the study of "genetic and familial" patterns. This last category has shown consistent growth since 1985.

Cycles of interest and credibility are common in biomedical research.[3] How should we look at these two waves? Optimistic observers, usually biologists, explain them by pointing out the knowledge gathered employing molecular biology tools and concepts. Though family pedigrees analyzed in 1932 and 1992 look similar, they show important differences since recent versions derive from the use of DNA probes, which make visible such predisposition factors as genome sequences. In contrast, critically minded commentators, usually sociologists, look at the social and cultural interests underlying genetic discourses and suggest that external constraints may account for the rush to identify cancer predisposition factors: in the 1930s, the eugenic movement; in the 1980s, the biotechnology industry. If the historian is to contribute to this deadlocked debate, presumably it is through a vision of change and continuity; in other words, in delivering an analysis of the material cultures that form the basis of the debates on human genetics.

This paper aims to follow one cycle of the *alternation* between viral, genetic, and chemical approaches to cancer. I shall focus on the mouse model for breast cancer since a glance at *Index Medicus* from 1930 to 1956 shows the existence of two difference regimes of publication. In the late 1930s, the vitality of the field of "cancer heredity" was reflected by a wave of interest in the inheritance of mouse mammary tumors. From 1936 to 1942, the *Index* mention more than 20 papers each year. Conversely, in the 1950s, a few papers following isolated surveys of family histories were published, and the mouse mammary tumor system was redefined as a model of virus action.

Among the many factors that enhance this alternation, one stems from the current emphasis on laboratory-based medical knowledge, namely the articulation between experimental animal models, (pre)clinical studies, and medical demands in general. Animal models of human disease make problematic tools since they cannot usually simultaneously be used as agencies in different material and social worlds and be standardized in order to generate unambiguous meanings. As *animal* models of *human* diseases, they are epistemologically underdetermined entities. Evaluation within different contexts results in processes akin to Collin's experimenter regress.[4] In order to confer value on the model, scientists and physicians must negotiate an agreement dealing at the same time with the properties of the human disease, the properties of the biological system, and the proper way to translate the latter into the former. Such models are therefore rarely compelling, since what is a convincing match is a highly variable and open issue that may generate rapid changes and alternations.

Once the variability and amenability of animal models is acknowledged, one is left with the problem of their long-term uses. One may stress that animal models mediate interactions between actors engaged in different "social worlds."[5] Thus animal models articulate the practices prevailing in the laboratory and the clinic, and, like many boundary objects, they are instrumental by virtue of their openness.[6] This overarching vision may explain how animal models are chosen to match existing commitments, but it does not say much about how they change and how usage changes the cultures and interests of interacting parties.[7] The contention adopted here is that animal models are stable and influential as long as standards can be made and circulated between different settings.[8] From a practical viewpoint, animal systems are valuable because they are workable: in contrast to human beings, time and resources can be invested in order to narrow down variability. Therefore, the development of biomedical models bears some similarities to production processes. It is a practice of homogeneity encouraged by the making of "interchangeable parts," and the integration of "routinized sequences of work."

Accordingly, the shift from genetic predisposition factors to vertically transmitted viruses that took place in breast cancer research after World War II reflects the existence of two regimes of practice. Up to the 1940s, the study of mouse mammary tumors was rooted in the production of inbred strains of mice. The process focused on the circulation of mice and the definition of protocols that could be employed as gauges in order to evaluate and homogenize local experiments. As such, it echoed the practice of industrial standardization.[9] In the late 1940s, cancer researchers tried to supersede an increasing variability of mouse mammary tumors. They redefined mouse models as instances of vertical infection. At the National Cancer Institute, they established large-scale screening operations aimed at the discovery of mouse tumor viruses through systematic "trial and error" processes. The new regime was informed by the post-Taylorian principles of large-system management that emerged from the war experience. Examining the uses of the mouse mammary tumor model will shed some light on the transition.

1. Clarence Cook Little's Genetic Equation

In 1937, after the National Cancer Act creating the National Cancer Institute was passed by Congress, Surgeon General Parran, then head of the Public Health Service nominated the first National Cancer Advisory Council (NCAC) and commissioned a report on opportunities in cancer research. The geneticist Clarence C. Little took part in this work. Little was a prominent figure in cancer research, the managing director of the American Society for the Control of Cancer, as well as the director of the Jackson Memorial Laboratory, a combined private research center and mouse production center. In the spring of 1937, during the press campaign that preceded the congressional debates on the Cancer Act, both *Time* and *Life* focused special issues on Little, his mice, and cancer research.[10] It is no surprise that the NCAC report, published a few months later, stated that two classes of etiological factors had to be taken into account: inherited susceptibility and carcinogenic chemicals.[11] Yet the text mentioned a new agent that seemed to be transmitted by the milk of suckling females and might induce tumors in mice in the same way as the action of chemicals. This cautious strategy for defining the causes of mouse mammary tumors reflected tensions that were overwhelming the Jackson Laboratories. Little and his staff were struggling then to run the laboratory along the lines established at its creation in 1929, namely, the study of cancer genetics by using genetically homogenous inbred lines of mice.

Little's interest in the production and selection of mice as tools for the study of cancer may be traced back to when he was a graduate student at Harvard, finishing a doctorate under the supervision of W.E. Castle.[12] Afterward, Little collaborated with E.E. Tyzzer, a pathologist who was working at Harvard Medical School on tumor immunology. Tyzzer had been investigating the variable fate of tumor grafts in mice, and he had come to the conclusion that susceptibility to transplanted tumors depended on the biological identities of both host and donor. While working on the part played by hereditary constitution, he obtained complex results that could not be classified on the basis of simple patterns. Knowing of Little's work on the complex inheritance of mouse color factors, he recruited the young geneticist. In Castle's laboratory, Little had set up a breeding program aimed at the establishment of strains of mice with stable color patterns, which could be used in hybridization experiments following the exemplar of Mendel's crossings. Little suggested using the same tools, i.e., strains inbred for a long time, to homogenize tumor transplantation and determine how many "Mendelizing" factors might control susceptibility.

Working with Little's "dilute brown" mice and with Tyzzer's Japanese waltzing mice, the two scientists achieved all-or-nothing experiments with a tumor growing in almost all mice from Tyzzer's stock and gradually disappearing in Little's stock. Moreover, the use of hybrids between the Japanese waltzing mice and others strains suggested that "resistance" to the graft was controlled by several independent, transmitted factors. A theoretical calculation suggested that something like ten dominant factors might control the implantation of tumor cells. The use of dilute brown was clearly viewed as an instrument whose main value was to narrow down the variability of transplantation experiments:

...the stocks used are genetically favorable for obtaining uniform and reliable experimental results. It seems important to emphasize this phase of the work, for if mixed or relatively impure races are used, variable and inconclusive results are almost certain to be obtained. We feel that the material used is of sufficient constancy and definiteness to lend strength to any experimental results obtained in the study of its hereditary behavior.[13]

From this case of transplanted tumors, Little later claimed that resistance to cancerous tissues was an inherited pattern controlled by several factors that obeyed Mendelian laws in mice and presumably in humans.[14] The collaboration of Little and Tyzzer at Harvard resulted in a significant displacement of the inheritance problem. Little and Tyzzer had viewed animals of known ancestry as necessary to achieve a control of the susceptibility to tumors. The local but successful routinization of transplantation experiments, achieved by using a single tumor type and "inbred" hosts, actually turned the homogeneity problem upside down. Transplantation patterns gradually became a means to assess the purity of the genetic background. Little started to employ transplantation to control the practice of inbreeding:

The race from which the animals used in the experiments were derived (the Japanese waltzing mice) is one that had been inbred for at least six years without any addition of new individuals from outside the stock. The result must therefore have been to produce a race of great uniformity with respect to whatever inheritable factors it may possess.... The homogeneity is further shown by uniformity of their reaction to their tumors....These results are in contrast with those obtained from the implantation of tumors in less inbred stocks....The uncertainty of the results in the experimental inoculation of the tumors of tame mice is a matter of common knowledge. Although the results obtained depend in part on the character of the individual tumor, this is to be regarded as a fairly constant factor, at least at any given transfer, so that presuming a satisfactory inoculation technic, constant uniformity in the growth of tumor implant furnishes strong evidence of racial homogeneity.[15]

After war service, Little resumed his work on coat and hair color in mammals. He came back to the study of cancer susceptibility temporarily while working at the Station for Experimental Evolution at Cold Spring Harbor. He started a new survey of resistance to grafts with the Japanese waltzing mice that led him to conclude that three to five factors were involved.[16] This work was interrupted, however, when Little embarked on an administrative career: in 1922, he accepted the presidency of the University of Maine. In fact, Little's reputation as a manager and reformer proved so successful that he was offered the presidency of the University of Michigan in 1925.[17] There he maintained a small research facility, where he kept a few lines, including his dilute-brown line (Dba) then inbred for more than ten years. Little was very active in the eugenics movement while in Michigan. He entered on a long-term collaboration with Margaret Sanger and became a strong advocate of birth control and women's suffrage. At the same time, he was involved in the management of the Eugenics Society. Pleading for genetic education and counseling, he saw mandatory measures such as sterilization as very crude ways of controlling the spread of bad genes that should be replaced by the use of contraceptive methods.[18] Speaking publicly against miscegenation, Little used to compare racial and laboratory purity:

I happen to be working in Maine where the population of the old New England stock is very, very high.... I don't want see that particular element in the situation mixed up, or mauled up. I want to keep it the way a chemist would prize a store of chemically pure substance that he wants to use for testing, that he wants to use for definite purposes when a certain element is needed.[19]

Little's speeches and reform initiatives stirred up such controversies at the University of Michigan that he finally resigned the presidency.[20] In 1929, he took on two new challenges. First, he accepted the nomination for managing director of the American Society for the Control of Cancer, a medical group dedicated to the education of physicians and lay people. Second, after securing the financial help of the Jackson and Ford families, he returned to Maine, where he wanted to establish a laboratory for the study of mouse genetics and cancer.[21] Little asked former collaborators to join him at Bar Harbor, among them an important recruit, Leonell C. Strong, a geneticist trained in Morgan's laboratory at Columbia. Strong had established several strains of mice with stable – either low or high – incidence of tumors. The Jackson Memorial Laboratory was established as a research setting, but it rapidly became an odd place where scientists were at the same time doing some research on mouse mammary tumors, producing inbred lines of mice, and linking eugenics to medicine.[22] A good illustration of the complex issues relating to the production system and the research program at the Jackson lab is the controversy that developed between Little and Maud Slye.

Slye was a former psychology teacher invited by Charles O. Whitman to work on heredity of behavioral traits at the University of Chicago.[23] In 1908, for unknown reasons, Slye gave up her work on the waltzing behavior of the Japanese waltzing mice and embarked on a study of cancer inheritance. The controversy with Little broke out when Slye published her first reports pleading for the existence of a single recessive factor governing cancer susceptibility. After a bitter exchange dealing with Mendel's law and the meaning of "dominant,"[24] Little let Slye's numerous reports go unchallenged. The controversy resumed in 1927, however, when the American Medical Association established a committee to examine Slye's proposal that an institute for the study of cancer inheritance in humans be established to generalize the data she had collected with mice.[25] Little was no longer an unknown scientist, and by 1928, he published a devastating article on Slye's work.[26] Technically speaking, Little used Slye's published data to assemble a statistical re-interpretation showing that her results did not accommodate a single recessive factor and that the inheritance of mammary tumor was of a dominant sex-linked type. The most interesting part of the controversy was concerned with the uses of inbred lines however. It echoed a previous argument about transplanted versus natural mice tumors. The debate was chiefly a matter of experimental practices, the Chicage "pedigree" culture opposing the would-be Bar Harbor "inbreeding" culture. After the Jackson Laboratory had been established, the debate developed into a methodological showcase for Jackson scientists without much reply on Slye's side.[27]

Slye's method of doing genetics was a mixture of breeding practices and anatomo-pathological studies. She was studying the incidence of cancer in mouse families, surveying the causes of death with systematic necropsies. She and her coworkers expanded the approach into an impressive system; between 1915 and 1928, they completed the analysis of some 60,000 mice.[28]

They gathered pedigree charts, drawings, and microscopic slides in a repository, the so-called "private Museum," associated with the mouse house. Her main argument against inbred lines was that the consistent practice of crossing between sisters and brothers resulted in weak and fragile animals leading to biased conclusions, since many of them were killed by infectious disease or other defects before any observation on cancer could be made. Moreover, the loss of fertility associated with inbreeding usually made it impossible to preserve strains beyond a few generations.[29] Finally, Slye opposed her "natural" (spontaneous) tumors to the laboratory artifacts employed by Little. She chose to stick to low degrees of consanguinity and to the comparison of pedigrees. Mice were inbred for a few generations in order to follow cancer inheritance, to enhance the birth of animals "homozygous for cancer" [sic], but pen breeding was kept in line in order to restore vigor. Slye was thus looking at cancer inheritance the same way as breeders selected traits such as coat color or size. Her use of the Mendelian framework was a way to describe transmission patterns within particular families, not a matter of statistics and computed ratios.

By contrast, the establishment of inbred lines at the Jackson Laboratory resulted in an attempt to impose order through the production of homogenous animals, which would circumvent the troublesome variability of "natural" cancer.[30] The Jackson research aimed at localizing the genes governing the formation of tumors with hybridization experiments (later linkage studies) based on strains showing stable but different incidence of a given tumor. The ideal inbreeding practice consisted in homogenizing the genetic pool with a series of crosses between brothers and sisters or between parents and offspring. After a few dozen generations, the animals were considered to be homozygous for most characters. Standardization did not aim at producing identical animals but at monitoring the part played by hereditary factors.[31] Compared with Slye's method, the Jackson inbreeding work showed similarities with the establishment of a production line. The process rested on changes of scale and the systematic organization of tasks in order to increase homogeneity of output. First, keeping a great number of animals was the only way to avoid the misfortunes that Slye predicted, to cope with the loss of resistance to diseases and injuries or the sudden "mutational" changes that can occur at any time within the stock. Second, a stringent organization of scientists and technicians was necessary to try to avoid errors of mating, bad marking, or uncontrolled behaviors such as mothers eating their litters. Environmental sources of variations should be controlled as far as possible: food, for instance, should be kept identical over the years. Third, the success of inbreeding could only be assessed through the gradual homogenization of phenotypes. Genetic homogeneity was assessed on the basis of routinized transplantation controls: all the animals within a given strain needed to show similar susceptibility to grafts. Any deviation from the standard pattern should lead the researcher to suspect expected but unpredictable events like mutations or unexpected but predictable events like management errors.

The climate of the 1930s reinforced these similarities with industrial standardization and Taylorism. In order to overcome the decreasing financial support of his sponsors, Little also stopped giving away his mice and started to compete with other mouse retailers. The uniformity of the mice became a matter of reliability and reputation in an emerging market. Little and his associates were campaigning for the systematic use of inbred lines and compared the new tools with normalized, standard chemicals introduced from the chemical industry.[32] Although much of the work aimed at keeping the existing stocks stable, variability was not necessarily an enemy. Sometimes, it could be used as a resource to create new mice. The engineering of mice was a mixture of pragmatism and discipline. For example, the females of mice "A" brought in by Strong died from mammary cancer but, in contrast to another high-cancer strain (C3H), virgin females were much less affected than breeding females.[33] This pattern resulted from a contingent achievement. Before 1927, Strong's selection process focused on the resistance to tumor transplantation. Then something happened and "several different albino mice developed spontaneous carcinoma" of the mammary gland. Attempts to modify the age at which the mice developed cancer or to increase the incidence of tumor virgin females were unsuccessful. The A strain was sold later as a tool to study the effect of breeding and hormonal influence on mammary tumor incidence.

The debate with Slye also sheds light on the relations between eugenics and cancer medicine. In eugenics textbooks, "proof" of the inheritance of cancer in mice has usually been attributed to Slye.[34] Both Slye and Little addressed medical and eugenics audiences, using cases of cancer running in human families and/or statistics of human cancer in connection with data collected in mice.[35] In the late 1920s, however, when Slye backed the notion that human cancer could be controlled in a few generations by enforcing eugenic regulation, Little adopted a new strategy. He started to play down strong claims about human heredity, echo the medical criticism of eugenics, and back a "new" human genetics.[36] The declining influence of the U.S. eugenics movement in the 1930s has been recognized widely by historians.[37] Rising concerns with economic issues, as well as hopes for a renewal of the government's role triggered by the Depression and the New Deal reinforced this trend. Interacting with physicians on a daily basis as

director of the ACSS, Little was well aware of this crisis.[38] Accordingly, he
increasingly used the animal analogy:

It may be pointed out in conclusion that the accepted method of making human
mating, viz. by controlled outcrossing combined with the inadequate records and
small numbers of progeny which commonly are encountered in human families,
militates against the practical use of controlled heredity as a means of reducing the
incidence of cancer in man. This, however, does not prevent the genetic approach to
the problem in the laboratory and the use of controlled homogeneous inbred strains
of mice from being extremely favorable material for pure scientific research into the
nature and causes of cancer.[39]

2. Mouser Network and Standardized Animals

At the Jackson Memorial Laboratory, the articulation of the inbred genetic
culture along with the new concern with medical issues centered rapidly
around the study of mouse mammary tumors. The process originated in
crossings targeted at undermining Slye's theory. By 1933, a paper signed
"Staff of the Jackson Memorial Laboratory" described inheritance of
maternal type in the progeny of reciprocal crosses between low- and high-
cancer strains.[40] Under the guidance of Little, Jackson scientists were
inclined to view this material effect as a case of cytoplasmic inheritance: the
female parent would contribute nonchromosomal genetic factor to the
offspring. Alternative explanations could be advanced nevertheless, and one
staff member, John Bittner, started an investigation of the role of nursing. In
1936, Bittner presented preliminary results suggesting that a few females of
the strain A showing a mammary tumor incidence of 88% could be
transformed into "low-tumor" animals with a "low-tumor" progeny
(showing a mammary tumor incidence of 25%) if nursed by "low-tumor"
mothers. The "extra-chromosomal influence" was transformed into a "*milk-
influence.*" Something governing the formation of mammary tumors was
present in the milk of nursing mothers since the transmission could be
discontinued by foster nursing.

  This behavior was not very satisfactory. It jeopardized eu/gen/et/ic
thinking and the whole prospects of producing inbred "cancerous" and
"noncancerous" strains. On the one hand, it reinforced the argument about
the complexity of cancer transmission.[41] On the other hand, it could be
turned into a case for some cancer contagion, which Little had always
opposed.[42] Bittner carefully avoided any claim about the nature of the
influence. He crafted a purely operational definition of the agency: it was a

milk influence enhancing the incidence of mammary tumor in suckling newborns. The transmitted influence nonetheless disturbed the local alchemy. First, the milk influence seemed to be "six to ten times" more important than the hereditary influence.[43] Second, if other high-incidence strains displayed similar influences, the new entities could render obsolete all the debates about inherited susceptibility to tumors, as well as the enterprise of genetic standardization for, as Oberling put it, "the science of genetics may be collapsing before our eyes."[44] Third, it came "naturally to the mind" that the influence might be a virus. The viral theory of cancer had limited impact within the cancer community, but it thrived in a few settings, particularly in William Gye's laboratory at the Imperial Cancer Research Fund in London.[45] Finally, even at the Jackson Laboratories, the milk influence looked like a "colloid of high molecular weight": it was active in dried tissue, and it was not retained by filters.[46]

In Bar Harbor, several paths to domesticate the agent and preserve mammary cancer genetics were followed. Little tried to circumvent the problem by reducing the milk factor to an endocrine agent or some form of physio-chemical influence of maternal tissues. The French biologist A. Lacassagne had announced, in 1932, that the injection of estrogen hormones over an extended period of time could induce mammary tumors in male mice, and his claim had triggered a wave of studies dealing with the effects of sex hormones on mouse tumors.[47] Accordingly, Little tried to link the milk influence with hormones by showing that blood of high-cancer strains could also induce the formation of mammary tumors in low-cancer strains.[48] Along the same line, he suggested the existence of some placental influence.[49] Commenting on the issue in 1944, Little presented a scheme based on "six vehicles of influences," from the transmission of chromosomes to nursing, that may govern the constitution of individuals.[50]

Bittner developed a package of practices and standards embodying a milk agent designed according to the "inbred" mice culture. A major caveat stemming from the foster-nursing experiments was that some factor might be transmitted vertically in all high-cancer strains, not only in strain A. Failure to reduce the incidence of mammary tumors by foster nursing meant that the homogeneity of tumor incidence, which had been viewed as a result of consistent inbreeding could originate in the standardization of husbandry, in identical environmental and maternal influences. The endless debate about nature and nurture was knocking at the door of the mouse house, and no decisive experiment could settle the matter. What was needed was a system of shared practices, a system of standards defining mice with high genetic susceptibility to tumors. The discussion about tumor transplantation had been managed precisely that way. Once the role of genetic constitution in graft rejection was recognized, the rejection of tumors in a given stock was used increasingly as a test for genetic heterogeneity, whose results "showed" that transplantation was genetically determined. Bittner's first move was to create a similar test for the presence of the milk agent. First, he proposed to establish susceptible strains. Lines with low incidence but showing high tumor rates after being nursed by mice from a high-tumor incidence would be considered as "susceptible but agent-free." Strains of unknown status would then be tested by the nursing of newborns from the susceptible line. One caveat was that the reagent line could show a latent or a temporarily inactive influence.[51] If necessary, multiple trials of the same sort could be completed. Beyond local adjustments, the circulation of reference animals made the case.

A few years after Bittner's experiment, foster nursing was actually routine practice, showing the presence of the milk influence in a half dozen laboratories. The diffusion pattern matches the circulation of reference mice: at laboratories working on the mammary tumor system, biologists established new colonies of C3H mice showing the milk influence and high tumor incidence.[52] This pattern suggests that the generalization of the *milk influence* bears some relationship to the circulation of inbred mice and foster nursing. Ready-made mice and protocols facilitated convergence among laboratories. The circulation of organisms was at once the circulation of a system, including research practices, and of problems worthy of investigation. The process led to the development of similar experimental strategies. Nonetheless, consensus did not emerge from straightforward use of homogenous resources; it originated in the tinkering with and the fine-tuning of local systems that were highly dependent on the distribution of credibility within the research community.

The development of milk-factor studies at the National Cancer Institute illustrates the process well. Work started in 1939, when the oncologists of the Public Health Service moved to their new laboratories in Bethesda.[53] H.B. Andervont, then working both on tumor viruses and tumor transplantation within the Harvard Public Health Service group for the study of cancer, began reciprocal foster-nursing experiments. Using his own colony of C3H mice, he tried to convert C 57 mice, which develop mammary tumor infrequently, into a permanent high-cancer strain. In 1940, Andervont and his group claimed a cancer rate of 14% (14 mice), whereas Bittner had already claimed 38% (8 mice) using the same C 57 line fostered by strain A mice.[54] Replication was nevertheless granted since there was a "significant increase of cancer incidence." Moreover, in the same issue of the *Journal of the National Cancer Institute* few weeks later, Bittner was realigning his results: with 104 fostered mice, the figure dropped to 11%.[55]

Then the scale of the operation was extended with a continued series of foster-nursing experiments going on with strain C3H at NCI and with strain A at Bar Harbor. In 1943, Andervont reported a surprising 63% (35 fostered mice), which was unexplained but welcomed.[56] Bittner kept on providing low but stable rates (18% with 98 mice in 1944), but at NCI, rates remained unstable (in 1948, Andervont reported a complete failure to increase the tumor incidence). Eventually, tests were interrupted, and the notion that the C 57 stocks used at NCI were genetically heterogeneous emerged as a convenient consensus since the nursing mothers (A and C3H) were established, standardized lines for cancer incidence. There seemed to be no point in checking uncomfortable hypotheses about the variability of the agent.

Similar problems were not rare. For a few years, however, the circulation of inbred strains originating in Bar Harbor provided a layer of common activities that was strong enough to enable variable adjustments within the framework provided by Jackson's scientists. One example was the sudden "appearance" of a tumor agent in mice from Bittner's stocks that had never been nursed by females with the agent, nor injected with extracts. The case was reported as a mutation, modifying the susceptibility to an inactive agent.[57] The phenomenon was later echoed when Andervont and Heston (a new NCI fellow trained at Jackson) backed the existence of different genes that determined the transmission and propagation of the agent. They were interpreting surprising changes of incidence in the progeny of backcross achieved with hybrids between C 57 and C3H mice.[58]

Bittner's second path to domesticate the milk influence was an attempt to disentangle the different agents. Once again, tumor transplantation provided the practical model. The "genetic" classification of tumors and mice had been worked out with trials in which the same tumor was used in different strains, or, alternatively, in which different tumors were tested against an accepted genetic reference, e.g., a single inbred line. Eager to establish similar crossings between the inbred lines and the milk influence, Bittner tried to compare the foster nursing of mice from lines "with" and "without" the influence. A third factor was introduced later with the injection of various hormones. The "inbred line" style was thus extended into an operational definition of the new factors. According to the status of nursing mothers, the mice would have an "active milk influence" or an "inactive milk influence." According to the number of litters or the injections of hormones, they would have "estrogen stimulation" or not.[59] A *triangulation* based on an independent assessment of the three factors was, impossible to achieve, however. Foster nursing might change both the hormonal influence and the milk influence, whereas the genetic constitution of high and low

strain might be a matter of different milk influences or of different
endogenous hormonal patterns. Since a clearcut hierarchy was almost
impossible to establish, Bittner crafted a theory based on the *interplay* of
three agencies:

(a) an active mammary tumor milk influence which is generally transferred by
nursing; (b) hormonal stimulation of the mammary tissue resulting in growth
suitable for cancerous change; (c) an inherited susceptibility to the development of
spontaneous mammary tumors, which may be transmitted by males and females of
the susceptible stock.[59]

Bittner's proposal was a loose explanation, but as such it was a remarkable
link between the local mousers and the diverse audiences interested in
mammary tumors. As a link between laboratory practices and medical
demands, it proved more successful than a theory of pure genetic
predispositions.[60] A major reason for the success of Bittner's three-way
model was that it encouraged mediations between mousers,
endocrinologists, and clinicians. In the late 1930s and early 1940s, the
industrial production of synthetic analogs of sex hormones led clinicians to
pay increasing attention to the use of hormone in breast cancer treatment.[61]
Following the emergence of the concept of a milk agent, a collective
evaluation of the mouse model proposed by the staff of the Jackson
Memorial Laboratory stressed therapeutic applications with a new tone that
opposed the concerns of applied eugenics:

As stated before, it is entirely possible that the practical procedure in therapy may
be able to rely more upon natural or artificial changes in internal secretions than
upon genetic constitutional differences. It would seem, however, safe to predict that
the utilization of known genetic strains of mammals which differ in their genetic
constitutional tendency to form mammary tumors will be one of the most important
means by which the influence of internal secretions will be accurately measured.[62]

Therefore researchers believed, hormones should align the mouse model
with a series of preclinical and clinical investigations. In Bar Harbor, the
association was rooted in a reorganization of work stemming from the
possible relations between the milk-influence and sexual hormones. It led to
preliminary attempts to cure mammary tumors following injection of
estrogens. Preclinical studies were not expanded in Bar Harbor, but they
thrived at the National Cancer Institute. For example, Andervont's strategy
consisted in an alternation between the different aspects of the mammary
tumor model. The milk agent was connected to a variety of studies,
including the effects of ovariectomy and castration, of subcutaneous

stilbestrol pellets, and of breeding.[63] Although these activities could be related to epidemiological studies in women or to clinical trials with sex hormones, experimental work on the mouse did not provide new information that could be transferred to the hospital ward. The research provided a useful interface, however, between two professional worlds, which could be used both by clinicians and tumor biologists. In the 1940s, many papers presenting clinical results started with a reference to Lacassagne, Bittner, and mouse mammary tumors.[64] In addition to the putative role of hormones in breast cancer (which could be traced by reevaluating statistics about age or pregnancy) the milk influence paradoxically encouraged a wave of studies of inherited disposition factors. In contrast to the pedigrees collected by the first group of studies in the 1920s that usually mixed breast cancer with other human tumors, the former became a particular topic of inquiry. Since constrained systems and predictive theories hardly fit the epidemiologist practice, the richer the model, the better. Thus Bittner's scheme and Jackson's mice could provide additional legitimacy to statistical surveys and *vice versa*.[65]

3. The Postwar Biomedical Complex and Incurable Variability: Turning Genetic Influences into Viruses

By the late 1940s, the conjunction of interests that had stabilized the mouse mammary tumor model waned. Two factors caused that to happen: first, Bittner's phenomenon was redefined as a case of virally induced mammary tumor; second, leukemia became the main topic of interest in experimental cancer research. There are several reasons for these changes. First, Little's personal equation suffered from the backlash against eugenics that followed World War II as well as from the reorganization of the American Society for the Control of Cancer. Second, the network of "mousers" expanded, and Jackson was no longer unchallenged as a leading center. The effect of standardized mice became less and less compelling. Third, the reorganization of biomedical research rooted in the war experience favored new patterns of research management, as well as new topics, especially at the National Cancer Institute.

Little's declining prestige within cancer research community in contrast to the genetic community in the United States calls for a few comments. A serious sign was the reorganization of the American Society for the Control of Cancer, which resulted in the creation of the American Cancer Society (ACS).[66] Mary and Albert Lasker got involved in cancer research in the early 1940s, and they did not follow the usual practice of sponsorship. They did more than simply create a foundation based on an endowment; they also joined the ASCC in 1942 and pressed for a new agenda that focused on research. Inherent in their approach was an increase in the influence of laypeople and the transfer of management skills from the business world. In the 1930s, Little accepted the physician's vision of an *educating*, sponsoring society, teaching the public how to think about cancer in order to enhance early detection.[67] By contrast, Mary Lasker and her allies introduced a sense of public relations and emergency reminiscent of New Deal activism. First, they organized regular and well-advertised fundraising events. Second, building on the experience of the "campaign" leading to the creation of the National Cancer Institute in 1937, they advanced a new strategy: lobbying the federal government for research support. Their marketing initiatives were successful and yielded Mary Lasker and her allies enough legitimacy to press for a reorganization. The American Cancer Society was born in 1943, and Little resigned his position of managing director.

A second reason for Little's declining role originates in the distrust of eugenicists that grew out of reactions against Nazi practices. By the late 1930s, American eugenicists were speaking of "democratic eugenics".[68] Nonetheless, in 1940, the Carnegie Institute closed its Eugenics Record Office in Cold Spring Harbor and, by the end of the war, the notion that human heredity should or could be put under control had become problematic. Similarly, a medically oriented American Society of Human Genetics took over the American Eugenics Society in 1948.[69] Concerns focused on a few diseases showing a simple Mendelien transmission pattern rather than complex pathologies. The idea of the heredity of human cancer quietly left the scene. For example, it disappeared from ACS material for public education.[70] The plea for "Research into the Biology of Human Cancer," actually a study of the factors governing individual "constitution," which Little made to the National Advisory Cancer Council, had almost no consequences.[71]

These changes reinforced emerging patterns at the Bar Harbor laboratory. In the 1930s, routine work had been managed by scientists, each of them organizing a particular branch housing his or her own colonies nurtured by a few technicians, animal breeders, and caretakers.[72] External demands triggered a partial reorganization with a new section and new colonies established with the single purpose of producing for the market.[73] Already in the late 1930s, Jackson Memorial was a place where hundreds of thousands of animals were produced every year. Systematic search for mutations creating new models of putative medical value was associated then with programs on transplantation and cancer inheritance.[74] More incidental events played a part too. After Strong left – he took a teaching position in New Haven – Bittner left in 1942 to head the division of cancer biology at the University of Minnesota Medical School.[75] Following their departure, the work dealing with the mouse mammary tumor system declined.[76]

Significantly, Bittner's move away from the Jackson environment was followed by a marked change in his account of the *milk influence*. In 1944, he reconsidered centrifugation experiments to conclude that the agent was not a colloid but "probably a virus."[77] Bittner did not revise the three-factor approach.[78] Further steps toward translation of the agent into a full virus were, made however, by focusing on cooperation with local biochemists.[79] Ultracentrifugation became the usual procedure for producing a cellular fraction containing microsomes and showing some mammary-tumor-inducing activity.

…the mammary tumor milk agent has the characteristics of an infectious agent or "virus" since:

1. It increases in amount in the presence of living cells.
2. It is macromolecular in size as shown by centrifugation.
3. It is specifically antigenic.
4. It remains active following lyophilization, desiccation, and filtration.
5. It appears to have a rather wide range of pH stability.[80]

Late testimonies suggest that Bittner adopted this position once freed of the sponsorship of Little, who strongly argued against any hint of viruses.[81] It should be noted that interest in the Rous sarcoma agent as a virus causing tumors reemerged at the same time.[82] A would-be molecular virology was then providing new resources for biomedical scientists: the model of bacteriophage and new instruments such as the ultracentrifuge and the electron microscope. Thus Bittner's move was not isolated. In 1945, the staff of the National Cancer Institute celebrated the transition to a new era of biomedical research with a symposia on mammary tumors in mice.[83] The event echoed the work done in the laboratory of Andervont, who stressed the similarity of the milk influence to Rous tumor virus.[84]

The redefinition of the milk influence as a virus could have had little impact.[87] However, the fate of the model was deeply altered by World War II. A glance at the articles published in the Journal of the National Cancer Institute show that mouse mammary tumors were on a decreasing course in the late 1940s and 1950s (see Table I). In contrast, the number of articles dealing with leukemia in mice increased rapidly after 1950. A computation of five-year means shows that there were two waves of leukemia studies, one

Table 1

Articles on mouse mammary tumors and mouse leukemia published in *the Journal of the National Cancer Institute.*

| Years | Mouse Mammary Tumors | | Leukemia | |
	Number	Annual mean	Number	Annual mean
1940-1944	29	5.8	4	0.8
1945-1949	21	4.2	7	1.4
1950-1954	20	4.0	25	5.0
1955-1959	28	5.6	31	6.2
1960-1964	43	8.6	65	13
1965-1968	33	8.2	70	17.5

starting in 1951, another in 1960, and that a moderate increase in the number of articles on mammary tumors occurred in the early 1960s. From this chronology and a qualitative survey of the articles one can conclude that:

a) until the late 1950s mouse mammary tumor studies focused either on the transmission of the milk agent or on the role of hormones in cancer; in the early 1960s, there was a renewed interest in the virus itself.

b) the study of mouse leukemia parallels the development of chemotherapeutic research at the NCI; the first wave was concerned chiefly with the testing of putative drugs against leukemia, the second wave with the same topic *plus* the study of mouse leukemia viruses.

This suggests that the NCI chemotherapy program, which focused on leukemia, triggered interest in mouse leukemia viruses. Later, the increasing commitment to tumor virus studies renewed interest in mouse mammary tumors, which were waning in the 1950s.

Before coming to terms with the impact of cancer chemotherapy, we must account for the declining prestige of the mouse mammary tumor model. The major blow seems to have been the lack of articulation of local systems. Up to World War II, the Jackson Laboratories had been the one reference center providing both animal tools and major results that enhanced the value of the model. In the early 1950s, lines of cancer mice showing mammary tumor viruses had been established in a dozen laboratories. At least seven important groups participated in the production of the results.[86] Paradoxically, the expansion of the mouser network produced a decentralized network that lacked authority.

Bittner's presentation of inbred lines is evocative of the changes enhanced by this delocalization. In the late 1940s, Little's emphasis on the purity of chemicals, purity of lines and purity of race had been replaced by an emphasis on *enigmas and pitfalls*. Bittner was still building on the analogy with the chemicals produced by industry.[87] Yet the stability and standardized patterns associated with inbreeding were no longer prevailing:

Whereas chemicals are generally stable over a long period of time, mice, like all living forms, are subject to 'mutational' changes which may occur at any time regardless of the degree of inbreeding.

These mutations were injurious to the biologist since:

If these are recognized they might be compared to the synthesis of new chemicals, new 'tools' for research; if they go unnoticed, the experimental results may be distorted because of the changes in the genetic constitution of the experimental animals and, therefore, may become the basis for new theories which cannot be confirmed.

Of course, mutations had been both expected and exploited. There were increasing concerns about discrepancies among laboratories allegedly using the "same" strains and methods. I have already mentioned the problem with C57 mice foster nursed by C3H mothers, and it became worse after the war. By 1948, Andervont had once again produced surprising results suggesting that, in some cases, no cancer-inducing agent could be transmitted from C3H mothers to C57 newborns. Moreover, in his lab, crosses between C57 females and C3H males could produce high-tumor litters, a result that suggested male transmission of the agent. Finally, back crosses between these hybrids and the parent mice segregated mice with high-incidence progeny and mice with a low incidence. Andervont and Heston therefore claimed that new factors were governing the transmission of the milk agent. First, the local C57 line was presumably heterogeneous for an unknown reason, and some C57 mice were transmitting an inactive agent.[88] Second, the milk agent could also be transmitted by males.[89] Third, two sets of mice genes participated in mammary tumor: one controlled susceptibility, the other determined the propagation of the milk agent.[90]

An increasing number of sublines and local practices actually multiplied such troublesome events. Commenting on these discrepancies and the many local theories, Bittner emphasized the fact that even Andervont's C3H colony was a separate line that could not be controlled by other lines of C3H: "Sublines have been found to exist within several of our most highly inbred stocks and these genetic differences may influence the incidence of spontaneous cancer." Worse, replication could not help: "*That* criterion of a good experiment therefore is faulty: for, as long as the errors remain constant, their effects will reappear."[91]

Internal controls could monitor incommensurability and changes within alleged similar lines, but they would not unify practices or provide a general scheme of interpretation. Since a collective organization of experiments triggering the negotiation of standards did not emerge from the network, local systems remained the rule. As a symptom of the rising uneasiness with the mouse mammary tumor model, interpretations regularly fell back on environmental factors and individual management of the colonies. For example, in the years 1950-1954, in contrast to what happened in the laboratory of Andervont, who had provided the nucleus of mice, Heston's colony of C3H mice foster nursed by C57 females "free of the agent"[93] repeatedly showed a 25% incidence of mammary tumor. Attempts to detect some transmissible agent or to craft another line with a normal low incidence failed. The high incidence was finally considered to be the result of some high genetic susceptibility enhanced by diet and breeding conditions that differed from Andervont's. From the mid-1950s on, reviews on the causes of mammary tumor in mice had become series of experiments organized in chronological order.[93] J.R. Heller reviewed 20 years of research at NCI, and reported on it in 1957:

The National Cancer Institute actually ran a mouse dairy at one time in the course of numerous attempts to characterize the milk factor. Mammary neoplasms can occur in the absence of the Bittner agent. They seem to be scattered randomly through the pedigree, showing no tendency toward the familial distribution that characterizes the occurrence of breast cancers among stocks harboring the agent. Individual susceptibility plays a dominant role in production of mammary cancer insofar as the milk factor is concerned.... The agent appears to be extrinsic rather than an endogenous produce of cellular metabolism and possesses infectivity relationships similar to those demonstrated for microorganisms.... The staff of NCI published an exhaustive study on "Mammary Tumors in Mice" during 1945, which defined the complex interrelationships among hormonal, microbiological, and genetic influences, and to which little can be added at this time.[94]

Most accounts of the fate of the mouse mammary model in the 1950s stress the fact that the milk factor was a bad choice for tumor virologists since it could not be produced by infecting cells in vitro, and it could not be properly isolated. But the real issue is to understand why a viral model was not legitimate in the 1950s since it worked in the 1970s with similar techniques. In other words, one might wonder why NCI laboratories did not replace Jackson Laboratories as a source of legitimacy and standards. The answer may be partly that NCI biologists were not involved in disciplining mouse mammary tumors studies because they were too busy working on leukemia viruses.

To explain the importance of leukemia at NCI, one must take into account the development of chemotherapy. War contracts in medical research and, most important, the history of the penicillin program contributed to a special emphasis on highly organized and targeted research based on federal contracts. This approach followed examples from industrial research and focused on the transfer of management schemes from industrial settings.[95] The trend is exemplified after 1945 by the creation of several chemotherapy programs located in major cancer research institutes, which were heavily supported by the American Cancer Society. For example, at Sloan-Kettering in New York, chemotherapy was at the center of the offensive for treatment launched by director Cornelius P. Rhoads the former chief of the Medical Division of the Chemical Warfare Service.[96] Preliminary programs relied on limited clinical trials and the use of a few compounds such as synthetic steroids and purine compounds that blocked the metabolism of folic acid. Thus, in 1948, Sidney Farber of the Massachusetts General Hospital announced "temporary remissions in acute leukemia in children" obtained by treatment with aminopterin, a folic acid antagonist.[97] Highly publicized by the ACS, in which Farber was a leading influence, the case focused many interests, hopes, and concerns on leukemia in childhood.[98]

The scale of operations changed with the increasing role of the NCI and the commitment of Congress.[99] Chemotherapy started there in 1945 with a small screening program based on the use of mouse subcutaneous sarcoma S37, which had been kept by serial transplantation for more than 40 years.[100] Late in 1952, in contrast to the recommendation of the National Advisory Cancer Council, Congress was told by Mary Lasker, Sydney Farber, and Cornelius Rhoads that a national chemotherapy program was necessary. Using the success of penicillin as an example, they believed that a well-founded, planned, systematic, centralized, and well-targeted initiative in chemotherapy would provide significant results. Congress bought their argument and appropriated $1 million. In 1954, NCI launched a national cooperative program, which integrated the work in industry, screening laboratories, and cancer hospitals.[101] NCI scientists were involved as contractors and advisers of the Cancer Chemotherapy National Service Center (CCNSC). Appropriations were increased to $25 million in 1958, and the program expanded into a system of drug development based on contracts with pharmaceutical companies, the organization of mass testing for compounds with potential antitumor effects, and the coordination of clinical trials.[102] Both the culture of mobilization that had originated during the war and the integration of highly divided work in industrial research teams provided managerial models. The explicit aim of the new program was, in the words of one of its main organizers, Gordon Zubord, "to set up all the functions of a pharmaceutical house run by the NCI." The CCNSC head, Kenneth Endincott, explained that the problems of cancer chemotherapy would be solved "when industry-government cooperation will be as effective in the pharmaceutical area as it is in some of the defense areas."[103]

Chemotherapy was strongly dependent on the supply of mice. The screening of putative antitumor compounds required the mass consumption of mice since the first stage of assays involved assessing the effects on transplantable mouse cancers. Thus chemotherapy did not treat mice as an experimental system, a physiological unit, showing links between cancer causes and the formation of tumor, but as a standard reagent providing a convenient medium for the growth of tumors and the test of chemotherapeutics. The meaning of mice standardization was changed since the practice no longer aimed at the control of one biological factor, i.e, genes, in order to simplify the model, but at producing identical animals. The engineer's problem was to keep the organism constant, free of other disease, in order to inject or graft tissues and induce a massive and rapid formation of tumors. The issue was to mass produce strains that would be good recipients of selected tumors. Centralized management resulted in the choice of only three mouse tumors for routine screening: one form of leukemia and two subcutaneous tumors. The choice of animals to be produced was limited: Swiss, Dba, and C57 BL strains, as well as hybrids of C57 BL and Dba. In 1956, the NCI set up a committee that established standards for the commercial production of random-bred and inbred laboratory mice to be implemented by all the institutes suppliers.[104]

As the program developed, needs increased: "the magnitude of the screening operation has vastly increased the demand for inbred mice and is taxing the facilities of mouse breeders operating within the United States."[105] So the CCNSC developed a mouse production program funded by a special NCI grant. The goals of the program were explicitly formulated in industrial terms: its directors discussed the volume of input and output of the product, the problem of standardization, and quality controls. First conducted at the Jackson Laboratories, the mouse breeding program later was extended to commercial laboratories (Battelle, Texas Inbred, Simonsen Laboratories, and Pfitzer).[106] All the animals used in screening tests had to be supplied by producers accredited by the CCNSC, and all the demands for supplies of mice were processed through the CCNSC's Mammalian Genetics and Animal Production Section.[107] Screening activities were also delegated to commercial and semicommercial facilities: Wisconsin Alumni Research Foundation, Microbiological Associates, Hazelton Laboratories, and so forth.

A paradoxical effect of the rise of this chemotherapy program focused on leukemia was to enhance the study of tumor viruses. For some years, links between the Cancer Chemotherapy National Service Center and the NCI laboratories were tenuous. The NCI was neither doing clinical trials nor testing chemicals, a process chiefly contracted to private centers. Nonetheless, a few scientists such as Lloyd W. Law were at the same time using inbred strains of mice to mimic clinical trials of chemotherapeutics and to study leukemia. They both managed their own colonies and used the NCI mouse facility. The former had been established in the late 1940s from breeding pairs that had originated either in Andervont's laboratory or in the Jackson Memorial. Exchanges with the CCNSC increased when the NCI's operations expanded and leukemia came to be seen as a case of vertical infection – that is when Ludwig Gross's work was adopted by NCI biologists.

Physician and microbiologist on duty with the U.S. Army Medical Corps, Gross started a research project on cancer causation at the Veterans Administration Hospital in the Bronx in 1945.[108] According to his later account, he had almost no relationship with the "mouser" network except via J. Furth at Cornell, who provided him with a nucleus of leukemic mice of the strain Ak. He then obtained C3H mammary tumor mice from Bittner. One project was to induce leukemia in C3H. At the same time, Gross did endorse a radical vision of cancer as a viral disease.[109] For example, writing for obstetricians in 1949, he highlighted the implications of the mouse mammary tumor model allegedly showing a "vertical epidemic." Moreover, "vertical epidemic of cancer is not inheritance but a vertical transmission of an extraneous parasitic agent."[110] Mouse mammary tumors showed important similarities with communicable diseases such as tuberculosis or syphilis. Expanding on simplified presentations of foster-nursing experiments, Gross was claiming that inbreeding was contingent on circumstances in the passage of the milk agent. The cause of mammary tumor was genetic only to casual observers obsessed with pedigrees. So the study of tumor viruses did not always require inbred lines. Gross was one step ahead of Andervont and other believers with regard to the viral milk agent: instead of playing down other factors within the multifactorial theory, he was building a one-factor system rooted in the bacterial vision of contagious diseases. Gross then pointed out promises of prevention:

...it is possible to assume, however, that mammary carcinoma of mice does not represent a form of cancer different from breast cancer in other mammals. If this assumption is correct, the law of obligate communicability may in the future be established also for breast cancer in such animals as rats, rabbits, or dogs, or perhaps also in woman. Should such a possibility materialize, the eradication of human breast cancer may become feasible by the simple method of artificial feeding of infants born to mothers having a family history of tumors.

This perspective opened up a whole range of activities. On the one hand, laboratory work would aim at working out correlations between human breast cancer and a virus similar to the mouse agent. These correlation studies were based on electron microscopic surveys of breast milk. On the other hand, Gross's scheme was to associated the mammary tumor virus to other viruses of obvious medical interest. Leukemia was not a bad choice since some types of blood cell cancer (erythroleukemia) in mice had been attributed to viral agents. Two strains of mice showing high incidence of leukemia had been respectively selected first by J. Furth at Cornell University, then by J. Mac Dowell at Cold Spring Harbor. Gross received a breeding pair from the former.[111] From 1945 on, he inoculated extracts of various organs from leukemic mice into animals of the mammary-tumor strain C3H but for five years, he could not transfer the disease. By the early 1950s, however, Gross announced that a change in the inoculation technique – namely the use of newborn mice less than 24 hours old – enabled the transfer of leukemia with filtered extracts of leukemia tissues that contained neither cells nor bacteria. In other words, he could claim the discovery of a mouse leukemia agent.[112]

At the National Cancer Institute, Lloyd W. Law was especially interested in Gross's results since he was involved in both the screening of chemotherapeutic compounds and the biological study of mouse models of leukemia.[112] Law and his associates could not replicate the transfer.[114] Attempts to prepare and inoculate spleen extracts failed to increase the incidence of leukemia in test mice. Falling back on his results, Gross announced that he could actually divide his C3H mice into two groups: one was very susceptible to the agent, but the other failed to develop leukemia.[115] The pedigrees of these two groups differed: susceptible mice were C3H animals originating in a breeding pair received from Bittner, nonsusceptible mice originated in a pair received from NCI biologist Andervont. Presumably, Law's mice were of the same kind. The debate was leading to a complex assessment of local histories and putative mutations. The controversy could have achieved nothing but by 1958, NCI scientists had domesticated and developed leukemia viruses.

Gross later claimed that the controversy with the NCI workers ended when they bothered to replicate his experiments.[116] His account is a classical bacteriology story stressing the selection of more potent viruses and changes of virulence in microorganisms. It seems that researchers like J.

Furth actually attempted specific replication of Gross's procedures.[117] NCI scientists, however, did something else. They neither replicated nor opposed his work; they expanded their own system of mouse tumor viruses. According to the first report on inoculation experiments, Law and his colleagues did not use the same sublines as Gross.[118] Their C3H mice originated in NCI sublines, which had been separated from Andervont's for 15 years. Moreover, they tried to increase the yield of leukemia transfer by changing the genetic background. First, they used C3H from Figgee's laboratory, which showed high leukemia incidence. Second, they inoculated hybrids between C3H mice and Ak leukemic mice. In the discussion, the authors simply explained that inoculation did not increase the incidence of leukemia in 283 inoculated mice. They reported Gross's explanation with the comment that "the possibility of differences in response due to subline differences in the C3H strain remains to be investigated further". In contrast, they could "corroborate and extend" other findings of Gross, namely that some extracts of leukemic tissues could induce the formation of large solid tumors of salivary glands. Gross had reported separation of two viruses by centrifugation and filtration. In contrast, NCI workers could not correlate the formation of parotid tumors with a specific receiving strain, or a specific protocol.

Gross then tried to increase the incidence of leukemia by changing the agent. In 1957, he selected a more potent virus by successive passages in newborn mice.[119] NCI scientists "confirmed" Gross's latest results in the same vein: they did not report the use of his passage A virus but expanded their own systems. In 1958, Sarah Stewart employed local techniques in tissue culture to isolate and multiply a "mouse-leukemia-derived parotid tumor agent" that would become polyoma virus.[120] In 1959, John Moloney inoculated C3H newborns with extracts from the transplanted sarcoma S37 used in the chemotherapeutic screen. Quite unexpectedly, the extracts did not induce a sarcoma but a leukemia. Pursuing the viral track, Moloney isolated a leukemia agent, which was viewed as a latent virus transmitted with the transplanted tumors for almost 50 years.[121] These and similar achievements resulted rapidly in the isolation of a half dozen mouse leukemia viruses. This scaling-up was related to the chemotherapy program in three respects.

First, the chemotherapy program provided the political means for establishing virus studies on a large scale. Impulses for the creation of a tumor virus program was born out of the campaign for the cure of leukemia. In 1958, the Laboratory of Viral Oncology was established. In 1964, following the creation of the Leukemia Task Force, Congress made a special appropriation of $10 million for the establishment of a Special Virus

Leukemia Lymphoma Branch. In 1968, the Special Virus Cancer Program was launched to expand these targets to other tumors, especially solid tumors that resisted chemotherapy. Later this provided enough financial support to trace many candidates for the role of viral tumor inducers, including the renewed mouse mammary tumor viruses.

Second, NCI officials "passed" along the chemotherapy system of management to the virus study branch. Management principles followed the chemotherapy scheme with integrated investments aiming at a single target, i.e., the isolation of a virus that caused leukemia in humans to be used to produce a vaccine. A system of contracts managed by a few officials was to help implement short-term tasks: the isolation of mouse leukemia virus antigens, the screening of tissues from other mammals, or the search for viral particles in human leukemic tissues under the electron microscope. Although discussions about the reproducibility and interest of the new viruses were important,[123] the most innovative part of the system was the organization of immunological trials. In the late 1950s, electron microscopy was the usual instrument used to complete correlation studies that addressed issues such as contamination, specificity of host, or specificity of pathological effects. The prospect of finding human leukemia viruses encouraged immunological practices because work with animal systems would help establish reference reagents to assess the origins of putative candidates found in humans. Moreover such work would develop skills. Also, investments in cell culture opened up new possibilities since enough crude or partially purified extracts became available to do immunological work by injecting cell-free preparations, collecting antisera, and checking reactions with "viral antigens."

Finally, tumor virus research rested on procedures that beared strong similarities to the mass production/screening system established in the chemotherapy program. Thus the culture of standardization was extended to the study of leukemia viruses. It displaced the mammary tumor models because the process enhanced the use of mice from strains with stable incidence of mammary cancer. Initially, private contractors were involved in the rush toward leukemia viruses; later they were employed to produce mammary tumor viruses as well. For instance, from 1965 on, Meloy Laboratories, a company launched with NCI support, implemented a contract to "propagate, concentrate, distribute Mammary Tumor Virus, perform immunological and biological assays for detection and quantitation; develop methods for propagation and detection of MTV antigens; conduct studies on the control of neoplasia in the susceptible murine host by vaccination with inactivated virus."[124] The natural source of virus was the C3H strain.

By the early 1970s, the Virus Cancer Program was important enough to host many virus-causing tumors in mice, as well as in cats, rats, and monkeys. NCI officials could fall back on mouse mammary tumors and claim that:

Breast cancer is a leading cause of death among women. The finding of a virus resembling a type B RNA oncogenic virus of mice in the milk of a significant number of women from high-risk breast cancer families strongly suggests a possible viral etiology of this disease. A major effort of the VCP will be directed towards determining the relationship of viruses to human breast cancer.... MTV is the only available animal model system in which approaches to the study of viruses as a cause of breast cancer may be developed.[125]

Conclusion

In conclusion, I wish to come back to the alternation issue by considering a dialogue that took place in the course of an NCI press conference in 1971:

Question: Are these particles viruses?
Spiegelman: They are particles which are indistinguishable from others which we call viruses (Laughter). That's caution. You're free to call them viruses, but I have my colleagues to worry about...
Question: Would you make the general recommendation at this point that no woman nurse in the period in which you're...
Spiegelman: No, certainly not, no. Look, if a woman has a familial history of breast cancer in her family and if she shows virus particles and if she was my sister, I would tell her not to nurse the child.
Question: Dr Spiegelman, the publications we represent have a circulation of many millions. You are asking us to tell women to go out and get a test which is not available.
Spiegelman: No, I am not telling you to tell them...
Huebner: No let's put it a little different way. The point is, the evidence is that this type of virus, the B-type virus, does cause cancer of the breast in highly specific fashion in the mouse, and the point is that this is genetically regulated as well, and some mice get it and some don't, and we know which strains get it in the laboratory and which strains don't, and we could certainly advise these mice not to nurse their babies (laughter), but the problem is that the type of virus we have in humans...is one that we have really to show is responsible for cancer in human...
Spiegelman: Yes, I mean you cannot start a scare like this when we don't really know for sure that this virus particle is the causative agent. All we have is an analogy with an animal system...,"[126]

Thus we have a troublesome encounter between tumor virologists and the press. A common misinterpretation would be to complain about the hasty uses of bad science since human breast cancer viruses have never been found. The important issue in this dialogue is the definition of the animal model. Three questions were addressed at the same time: What is the mouse virus? What is human breast cancer? What about the uses of foster nursing? Each party tested what was reasonable to say and to expect at that time. The alternation between genetic and viral visions of cancer stemmed from hundreds of similar debates that defined both bench practices and their promises.

The redefinition of the mouse mammary tumor models and the shift from hereditary cancer factors to cancer viruses represent major trends in twentieth century health politics and culture. The contention underlying this paper is that the alternation between genes and viruses did not emerge from external forces shaping interpretations, but from a multi-layered articulation of practices. The history of the mouse mammary tumor model displays a succession of two regimes of experimentation. In the 1930s, a regime based on the production of inbred strains of mice was centered on the "mouser network" associated with the Jackson Memorial Laboratory. Researchers then envisioned an association between eugenics and medicine, but the lab's output was "constitutional influences" rather than "hereditary factors." A regime enhancing the study of mouse leukemia viruses emerged in the 1950s from the NCI's command of chemotherapeutic research. Relying on the management of large systems and the practice of mass production, this new regime focused on systematic screening operations whose aims were the discovery of cancer viruses and, later, the production of cancer vaccines.

The transition from one regime to the other originated not only in scientific and social trends that were the result of the war experience or the rise of virology but also in local conflicts related to the production of animal models. The spreading of Bittner's milk influence generated an increasing variability that proved unsolvable given the increasing number of mousers. In contrast, given the chemotherapy program, the displacement of Gross's leukemia virus to the National Cancer Institute resulted in an integrated system that produced many cancer viruses, as well as leukemia models. Tensions between homogeneity and variability, control and autonomy are inherent to the practice of standardization. The story of the mouse model of breast cancer shows that the stability of biomedical knowledge often depends on the existence of centralized authorities regulating the production and uses of experimental systems.

Acknowledgements

Part of this paper is based on collaborative research on the history of inbred mice initiated with Ilana Löwy. I am indebted to Ilana for many suggestions. The paper was completed while working at the Cambridge Wellcome Unit for the History of Medicine. I gratefully acknowledge the support of the Wellcome Trust. I thank Pnina Abir-Am, Angela Creager, and Daniel Kevles for their wise comments.

Notes and References

1. "Inherited breast cancer is a common genetic disease: 5% of a disease affecting one in ten women over the life span means that roughly 1 in 200 women will develop breast cancer by reason of inherited susceptibility. Therefore, as an inherited trait, breast cancer is one of the most common genetic diseases in the industrialized world." M.C. King. "Breast Cancer genes: How Many, Where and Who Are they?" *Nature Genetics, 2* (1992), pp. 89–90.
2. When coming to the issue of "eugenical applications," she condeded: "Unfortunately there seems to be little hope from the eugenic standpoint of eradication, for usually cancer does not develop in the patient until his children are mature.... It is different with tumors that develop in early childhood such as retinal glioma, which snatch a child from existence at an early age, or plunge him into a world of darkness for a life time. There eugenics has a definite role, and those who have retinal glioma and survived the operation should not pass on the defect to their children...." M.T. Macklin. "The Hereditary Factor in Human Neoplasms," *The Quarterly Review of Biology, VII* (1932), pp. 255–281.
3. P. Keating, A. Cambrosio, M. Mackenzie, "The Tools of the Discipline: Standards, Models and Measures in the Affinity/Avidity Controversy in Immunology" in *The Right Tools for the Job*, ed. A.E. Clarke and J. Fujimura (Princeton: Princeton University Press, 1992; and I. Löwy. "Experimental Systems and Clinical Practices: Tumor Immunology and Cancer Immunotherapy, 1895-1980" *Journal of the History of Biology, 27* (1994), pp. 403–435.
4. H. Collins, *Changing Order* (Beverly Hills, CA: Sage. 1985).
5. A. Strauss, "A Social World Perspective," *Studies in Symbolic Interaction, 1* (1978), pp. 119–128; and A. Strauss. "Social Worlds and Legitimation Processes," *Studies in Symbolic Interaction, 4* (1982), pp. 171–190.
6. S. Leigh Starr and J.R. Griesemer, "Institutional Ecology: Translation and Boundary Objects," *Social Studies of Science, 19* (1988), pp. 387–420.
7. For a discussion of this problem see Clarke and Fujimura, *The Right Tool.*
8. For exemplars of the stabilization of scientific agencies by the production of standards, see C. Smith and M.N. Wise, *Energy and Empire,* (Cambridge, England: Cambridge University Press, 1989); S. Schaffer, "Late Victorian Metrology and its Instrumentation: A Manufactory of Ohms" in *Invisible Connections: Instruments, Institutions and Science*, eds, R. Bud, S. Cozzens (Bellingham, WA: SPIE Optical Engineering Press, 1992); and J. O'Connel, "Metrology: The Creation of

Universality by the Circulation of Particulars," *Social Studies of Science, 23* (1993), pp. 129–173.

9. For a discussion of the production of animal systems, see Robert E. Koehler, "Systems of Production: Drosophila, Neurospora and Biochemical Genetics," *Historical Studies in the Biological and Physical Sciences, 21* (1991), pp. 87–127; R.E. Koehler, "Drosophila: A Life in the Laboratory," *The Journal of the History of Biology, 26* (1993), pp. 281–310; and B. Clause, "The Wistar Rat as a Right Choice: Establishing Mammalian Standards and the Ideal of Standardization," *The Journal of the History of Biology, 26* (1993), pp. 329–349.

10. "Pete Little was practically born a geneticist... At Bar Harbor, in a small building whose solid brick walls exclude stray mice, he produces 150,000 mice a year, sells 50,000 to other scientific institutions for research, anatomizes 25,000 to analyse their inherited characteristics, especially their susceptibility to cancer," *Time*, March 22, 1937, p. 54.

11. National Cancer Advisory Council, Report 1937, reprinted in *Journal of National Cancer Institute, 19* (1957).

12. Little's research career deserves an extensive treatment. His role in the development of inbred strains of mice is the topic of K. Rader's thesis: *Making Mice: C.C. Little, the Jackson Laboratory and the Standardization of Mus musculus for Research*. Indiana University, forthcoming. On his commitment toward eugenics, see R.G. Clark, *The Social Uses of Scientific Knowledge. Eugenics in the Career of C.C. Little* (University of Maine, 1956, Master's dissertation).

13. E.E. Tyzzer. "Tumor Immunity," *Journal of Cancer Research, 1* (1916), p. 125. C.C. Little and E.E. Tyzzer, "Further experimental Studies of the Inheritance of Susceptibility to a Transplantable Tumor, Carcinoma (J.W.A.) of the Japanese Waltzing Mouse," *Journal of Medical Research, 33* (1916), pp. 393–427.

14. C.C. Little, "The Relation of Heredity to Cancer in Man and Animals," *Science Monthly, 3* (1916), pp. 196–202.

15. Little, and Tyzzer, "*Further Experimental Studies...*, p. 396.

16. See C.C. Little, "The Heredity of Susceptibility to a Transplantable Sarcoma and (JWB) of the Japanese Waltzing Mouse," *Science, 51* (1920), pp. 467–468; C.C. Little and W.B. Johnson. "The Inheritance of Susceptibility to Implants of Splenic Tissue in Mice. I. Japanese Waltzing Mice, Albinos, and Their F1 Generation Hybrids," *Proceedings of the Society for Experimental Biology and Medicine, 19* (1922), pp. 163–167.

17. On this dimension, see Clark, chapter 4. "The Scientist as University President."

18. "It seems to me that the efforts toward sterilization which are now fairly widespread have in general a great deal of merit, but they are a rather crude implement. They are so to speak, biological prohibition, instead of biological temperance.... That is curative, not preventive, and sterilization does not reach and probably never can reach the type of individual who is a real menace to civilization, and at the same time one of its law-abiding and useful citizens.... I refer to an individual, normal himself, a good, law-abiding, constructive member of a community, but who carries the hereditary trait in his germ cells of epilepsy or feeble-mindedness or insanity." C.C. Little, "Unnatural Selection and its Resulting Obligations," *Birth Control Review. 10* (1926), pp. 243–244.

19. *Ibid.* p. 244.

20. See Clark, "*The Scientist as University President.*"

21. J. Holsten, The First Fifty Years at the Jackson Laboratory (Bar Harbor: The

Jackson Laboratory, 1979).

22. On the founding chart, see E.L. Green, "The Jackson Laboratory: A Center for Mammalian Genetics in the United States," *Journal of Heredity, 57* (1966), pp. 3–12.

23. For biographical information, See J.J. McCoy, *The Cancer Lady: Maud Slye and her Heredity Studies* (Nashville, TN: T. Nelson: 1977).

24. See C.C. Little "Cancer and Heredity" *Science, 42* (1915), pp. 218–219. C.C. Little, "The Inheritance of Cancer," *Science, 42* (1915), pp. 494–495; and M. Slye, "A Reply to Dr. Little," *Science, 42* (1915), pp. 246–248.

25. McCoy, *The Cancer Lady.*

26. C.C. Little, "Evidence That Cancer Is Not a Simple Mendelian Recessive," *Journal of Cancer Research, 12* (1928), pp. 30–46.

27. Many articles written by Little before World War II include critical views about Slye's recessive cancer genes. See for example C.C. Little, "The Role of Heredity in Determining the Incidence and Growth of Cancer," *American Journal of Cancer, 15* (1931), pp. 2780–2789; and C.C. Little, "The Present Status of Our Knowledge of Heredity and Cancer," *JAMA. 106* (1936), pp. 2234–2235. For the contribution of Little's collaborators, see: J.J. Bittner, "The Genetics of Cancer of Mice," *Quarterly Review of Biology, 13* (1938), pp. 51–64. Slye's reputation was severely damaged by Little's article. In 1936, she changed her theory, introducing two sets of genetic factors, one set controlling susceptibility to one cancer type, the other set the localization of the tumor. This change has been viewed (Shimkin, 1973) as a closure of the controversy since cancer was turned into a multigenic disease. The move, however, just displaced the arguments. The controversy closed when Slye retired and stopped publishing on cancer.

28. M. Slye, "The Relation of Heredity to Cancer," *The Journal of Cancer Research, 12* (1928), pp. 83–133.

29. This line of argument was already established when Slye published her first reports. See M. Slye, "The Incidence and Inheritability of Spontaneous Cancer in Mice," *The Journal of Cancer Research, 32* (1915), pp. 159–200.

30. See Green, *The Jackson Laboratory*; and H.C. Morse III, (ed), *Origins of Inbred Mice* (New York: Academic Press, 1978).

31. This contrasted with the making of other laboratory animals. See Clause, *"The Wistar Rat...".*

32. See C.C. Little, "The Present Status of the Cancer Problem," *Annals of Surgery, 93* (1931), pp. 11–15 and C.C. Little, "Some Contributions of the Laboratory Rodents to Our Understanding of Human Biology," *American Naturalist, 73* (1939), pp. 127–138. See also the volume produced collectively by the staff members, G.D. Snell, ed. *Biology of the Laboratory Mouse,* (New York: MacGraw-Hill, 1941).

33. L.C. Strong, "The Establishment of the A Strain of Inbred Mice," *Journal of Heredity, 28* (1937), pp. 21–24.

34. R.R. Gates, *Eugenics and Heredity,* chapter 14. (London:.Constable, 1929).

35. "The *fact* of inheritance is clear, but the *type* of inheritance needs further investigation. It does not appear to be simple Mendelian inheritance." C.C. Little, "The Relations of Genetics to the Problems of Cancer Research." *Harvey Lectures 1921–1922.* pp. 65–88. Slye's arguments are developed in Slye, *"The Relation of Heredity to Cancer."*

36. "The fact that the tendency to certain forms of cancer is hereditary in mice has been established for some years.... For a time the medical profession was not very

willing to accept this fact. Their experience had dealt with human beings, who had bred so slowly and had produced such a small number of progeny from any one mating, and were so difficult to examine and diagnose by the certain method of autopsy, that the details and method of inheritance were none too clear. It was therefore entirely natural that the profession did not at once agree with the conclusions of those engaged in laboratory research.... Too many false hopes have been built up and later destroyed in many investigations on cancer to allow a simple but incorrect genetic interpretation to go unchallenged, especially if it is not well supported by its proponent's data." C.C. Little, "Evidence That Cancer is Not a Simple Recessive", *Harvey Lectures 1921-1922*.

37. D. Kevles, *In the Name of Eugenics* (Berkeley: University of California Press, 1985); K.M. Ludmerer, *Genetics and the American Society*. (Baltimore: The Johns Hopkins University Press. 1972).
38. "There has been such a psychological and sociological maelstrom of ballyhoo and propaganda on economic and allied issues that the country is still spell bound or bilious.... Organized eugenics in the United States is in the doldrums because of several facts. Among these may be cited the following: (1) Lack of confidence on the part of experimental scientists in what has been termed eugenic research; (2) Lack of courage on the part of leaders in biological, sociological and economic fields in facing the conscious control of the quantity and quality of human population; (3) Inability of the average voter at present to see that many if not all our major ills of today are dependant on the fact that we have not used our intellect in the making of men as we have in the production of machinery." C.C. Little, "Not Dead but Sleeping," *The Journal of Heredity*, 24 (1933), pp. 149–150.
39. C. Little, "The Present Status of Our Knowledge of Heredity and Cancer," *JAMA.*, 106 (1936), pp. 2234–2235.
40. Staff of Roscoe B. Jackson Memorial Laboratory. "The Existence of Non-chromosomal Influence in the Incidence of Mammary Tumors in Mice." *Science*, 78 (1933), pp. 465–466.
41. W. Murray and C.C. Little "Extrachromosomal Influence in Relation to the Incidence of Mammary and Non-mammary Tumor in Mice," *American Journal of Cancer,* 27 (1936), pp. 516–518.
42. C.C. Little, "Education in Cancer," *American Journal of Cancer*, 15 (1931), pp. 280–283.
43. W.S. Murray and C.C. Little. "Chromosomal and Extrachromosomal Relation to the Incidence of Mammary Tumors in Mice," *American Journal of Cancer,* 37 (1939), pp. 536–552.
44. C. Oberling, *The Riddle of Cancer*, (New Haven: Yale University Press. 1944), p. 162.
45. J. Austoker, *A History of the Imperial Cancer Research Fund*, (Oxford: Oxford University Press, 1988).
46. J.J. Bittner, "The Milk Influence of Breast Tumors in Mice," *Science*, 95 (1942), pp. 462–463.
47. A. Laccassagne, "Relationship of Hormones and Mammary Adenocarcinoma in the Mouse," *American Journal of Cancer,* 37 (1939), pp. 414–424.
48. G.W. Woolley, L.W. Law, and C.C. Little "The Occurrence in Whole Blood of Material Influencing the Incidence of Mammary Carcinoma in Mice," *Cancer Research, 1*, 1941, pp. 955–956.
49. E. Fekete and C.C. Little, "Observations on the Mammary Tumor Incidence of

Mice Born from Transfered Ova," *Cancer Research, 2* (1942), pp. 525–530.

50. His series rested on three putative mechanisms: inheritance of chromosomes controlling rates of growth in cells, cytoplasmic particles controlling development and differentiation, hormones controlling the development of mammary glands. See Little. "Parental Influence on Cancer," *American Journal of Cancer, 15,* 1931, p. 97.

51. H.B. Andervont, "The Milk Influence in the Genesis of Mammary Tumors" in Staff of the National Cancer Institute, *A Symposium on Mammary Tumors*, Publication of the AAAS, Number 22, ed., F.R. Moulton, 1945.

52. The circulation of all Jackson mice showing mammary tumors do not reveal the same pattern. The fact that strain A, which was used in Bittner's original foster-nursing experiments, did not experience the fate of C3H is suggestive of the contingency of the circulation process.

53. J.B. Shimkin, "As Memory Serves: An Informal History of the National Cancer Institute, 1937–1957," *Journal of the National Cancer Institute, 59* (1977), pp. 559–600, see pp. 566–567 on the career of Andervont.

54. H.B. Andervont, "Influence of Foster Nursing Upon Incidence of Spontaneous Mammary Tumor Cancer in Resistant and Susceptible Mice," *Journal of the National Cancer Institute, 1* (1940), pp. 147–153.

55. J.J. Bittner, "Breast Cancer in Mice as Influenced by Nursing," *Journal of the National Cancer Institute, 1* (1940), pp. 155–168.

56. H.B. Andervont, "Influence of Hybridization Upon the Occurrence of Mammary Tumors in Mice," *Journal of the National Cancer Institute, 3* (1943), 359–365.

57. J.J. Bittner, "Changes of the Incidence of Mammary Carcinoma in Mice of the "A" Stock," *Cancer Research, 1* (1941), pp. 113–114.

58. W.E. Heston, M.K. Deringer, and H.B. Andervont. "Gene-Milk Agent Relationship in Mammary-Tumor Development," *Journal of the National Cancer Institute, 5* (1945), pp. 289–307.

59. J.J. Bittner, "Possible Relationship of the Estrogenic Hormones, Genetic Susceptibility, and Milk Influence in the Production of Mammary Cancer in mice," *Cancer Research, 2* (1942), pp. 710–721.

60. *Index Medicus* shows evidence of this success. From the late 1930s on, the milk agent was the main topic of a dozen articles a year.

61. A. Haddow, J.M. Watkinson, and E. Paterson, "Influence of Synthetic Oestrogens upon Advanced Malignant Disease," *British Medical Journal, 2* (1944), pp. 393–399.

62. Staff of the Roscoe B. Jackson Memorial Laboratory, "The Constitutional Factor in the Incidence of Mammary Tumors," *American Journal of Cancer, 27* (1936), pp. 551–555.

63. Review in M.B. Shimkin, "Hormones and Mammary Cancer in Mice" in Staff of the National Cancer Institute. *A Symposium on Mammary Tumors.*

64. S.G. Taylor *et al.* "The Effect of Sex Hormones on Advanced Carcinoma of the Breast," *Cancer, 1* (1948), pp. 604–617.

65. For instance, the British geneticist L. Penrose concluded one such study: "In view of the work on mice by Bittner, a factor derived from maternal cytoplasm, transmitted by way of milk, colostrom or cytoplasm of the ovum might be a specific cause. If this were so, maternal relatives should be more frequently affected with mammary cancer than the corresponding paternal relatives.... (Our) figures are suggestive though scarcely conclusive of maternal line inheritance." L.S. Penrose,

H.J. Mac Kenzie, and M.N. Karn, "A Genetic Study of Human Mammary Cancer," *British Journal of Cancer*, 2 (1948), pp. 168–176.

66. J.T. Patterson. *The Dread Disease: Cancer and Modern American Culture* (Cambridge: Harvard University Press, 1987) chapter 6; See also D.F. Shaughnessy, "The Story of the American Cancer Society," Columbia University, 1957, doctoral dissertation, chapter 12.

67. Shaughnessy, chapter 11. Patterson, chapter 7.

68. *The Journal of Heredity* then regularly hosted articles on the issue.

69. See Ludmerer, *Genetics and the American Society*, chapter 8; and Kevles, *In the Name of Eugenics*.

70. A survey of the ACS journal *Cancer News* for the years 1947–1955 shows *no* articles about cancer genetics in humans, although the Committee on Growth established jointly with the National Research Council in 1947 envisioned some work in the field.

71. C.C. Little, "Program for Research on the Biology of Human Cancer," *Journal of the National Cancer Institute*, 2 (1941), pp. 133–137. The first article on the genetics of human cancer published in the NCI journal came out in 1959. The author was an old-guard figure: M.T. Macklin.

72. E.S. Russel, "Origins and History of Mouse Inbred Strains: Contributions of C.C. Little," in *Origins of Inbred Mice*, ed. H.C. Morse III, pp. 33–44.

73. A sketchy description of the system can be found in *Mouse News Letter*, 5 (1951), pp. 24–36.

74. E.L. Green, "The Jackson Laboratory: A Center for Mammalian Genetics in the United States," *Journal of Heredity*, 57 (1965), pp. 3–12.

75. See L. Strong, "Inbred Mice in Science," in *Origins of Inbred Mice*, ed. H.C. Morse III, pp. 45–68; and C.C. Little, "J.J. Bittner," *Oncologia*, 16 (1963), pp 354–356.

76. A general review on mouse mammary tumors written by L. Dmochowski in 1953, mentions dozens of articles published by members of the staff of the Jackson Memorial Laboratory after 1944 and more than 70 during the years 1933–1943.

77. J.J. Bittner, "Observations on the Inherited Susceptibility to Spontaneous Mammary Cancer in Mice," *Cancer Research*, 4 (1944), pp. 159–167.

78. J.J. Bittner, "The Causes and Control of Mammary Cancer in Mice," *Harvey Lectures 1946–1947*, pp. 221–246.

79. C. Barnum, Z.B.B. Ball, and J.J. Bittner, "Some Properties of the Mammary Tumor Milk Agent," *Cancer Research*, 6 (1946), pp. 573–581.

80. J.J. Bittner, "Some Enigmas Associated with the Genesis of Mammary Cancer in Mice," *Cancer Research*, 8 (1948), pp. 625–639.

81. Strong, "*Inbred Mice in Science.*"

82. P. Rous, "The Nearer Causes of Cancer," *JAMA*, 122 (1943), pp 573–581; I. Löwy, "Variance of Meaning in Discovery Accounts: The Case of Contemporary Biology," *Historical Studies in the Physical and Biological Sciences, 21* (1990), pp. 87–121.

83. Staff of the national Cancer Institute, *A Symposium on Mammary Tumors*.

84. Andervont, "*Influence of Foster Nursing...*" p. 134.

85. The NCI, for instance, tried to collect mouse milk containing the agent on a large scale. H. Kahler, "Apparatus for Milking Mice," *The Journal of the National Cancer Institute*, 2 (1942), pp. 457–458.

86. Dmochowski's review of 1953 gives similar weight to the contribution of groups headed by Little (Jackson), Gross (New York), Korteweg (Amsterdam), Bittner

(Minneapolis), Heston (NCI), Andervont (NCI), Dmochowski (Leeds).

87. "Any advances in our knowledge about the genesis of mammary cancer in mice may be attributed to the availability of homozygous or inbred stocks. For research on these strains are to the biologist as pure chemicals are to the chemist." Bittner, *Some Enigmas.*

88. H.B. Andervont, and T.B. Dunn. "Mammary Tumors in Mice Presumably Free of Mammary Tumor Agent." *Journal of the National Cancer Institute, 8* (1945), pp. 227–233.

89. H.B. Andervont and T.B. Dunn, "Further Studies on the Relation of the Mammary-Tumor Agent to Mammary Tumors of Hybrid Mice," *Journal of the National Cancer Institute, 5* (1948), pp. 89–104.

90. W.E. Heston, M.K. Deringer, and H.B. Andervont, "Gene-Milk Agent Relationship in Mammary-Tumor Development," *Journal of the National Cancer Institute, 5* (1945), pp. 289–307.

91. Cannon quoted by Bittner, *Some Enigmas,* p. 627.

92. W.E. Heston, "Mammary-Tumor in Agent-Free Mice," *Annals of the New York Academy of Science, 71* (1958), pp. 931–942.

93. See L. Dmochowski, "The Milk-Agent and the Origin of Mammary Tumors in Mice," *Advances in Cancer Research 1* (1953), pp. 103–172, and J.J. Bittner, "Genetic Concepts in Mammary Cancer in Mice," *Annals of the New-York Academy of Science, 71* (1958), pp. 943–974.

94. J.R. Heller, "The National Cancer Institute: A Twenty-Year Retrospect," *Journal of the National Cancer Institute, 19* (1957), pp. 195 and p. 197.

95. R. Bud, "Strategy in American Cancer Research After World War II," *Social Studies of Science, 8* (1978), pp. 425–459.

96. Bud, "*Strategy in American Cancer Research...*", C.P. Rhoads, "Nitrogen Mustards in the Treatment of Neoplastic Disease," *JAMA, 131* (1946), pp. 656–658.

97. S. Farber *et al.* "Temporary Remissions in Acute Leukemia in Children Produced by Folic-Acid Antagonist, 4-aminopteryl-glutamic acid (aminopterin)," *New England Journal of Medicine, 238* (1948), pp. 787–793.

98. J.T. Patterson, *The Dread Disease* chapter 7.

99. NCI appropriations experienced an exponential growth then: $0.6 million in 1945, $14 million in 1948, $48 million in 1957.

100. C.G. Zubord. "Origins and Development of Chemotherapy Research at the National Cancer Institute," *Cancer Treatment Reports, 68* (1984), pp. 9–19.

101. See The National Program of Cancer Chemotherapy Research, *Cancer Chemotherapy Reports,* January 1959; and I. Löwy, *Between Bench and Bedside* (Cambridge: Harvard University Press, 1996).

102. See K.M. Endicott, "The Chemotherapy Program," *Journal of the National Cancer Institute, 19* (1957), pp. 257–293, and Löwy, "Nothing More to Do...."

103. Zubord, "*Origins and Development of Chemotherapy...*", p. 12; Endicott, '*The Chemotherapy Program...*', p. 292.

104. See B.F. Hill, "Laboratory Animal Standardization in the United States of America" in *Notes for Breeders of Common Laboratory Animals,* ed. G. Porter and W. Lane-Petter, (London: Academic Press, 1962).

105. Endicott, "*The Chemotherapy Program,*" p. 281.

106. Endicott, "The Chemotherapy Program," p. 280; C.G. Zubord, S. Schepartz, J. Leiter, K.M. Endicott, L.M. Carrese, and C.G. Baker, "History of the Cancer Chemotherapy Program," *Cancer Chemotherapy Reports, 50* (1966), p. 364.

107. Zubord, Schepartz, and Carter, "Historical Background of the National Cancer Institute's Drug Development Trust," *National Cancer Institute Monographs, 45* (1975), p. 10.
108. L. Gross, M. Bessis, interview "How the Mouse Leukemia Virus Was Discovered," *Nouvelle Revue Française d'Hématologie, 16* (1976), pp. 287–304.
109. This perspective may be traced back to his work on tumor immunology at the Pasteur Institute in Paris. A. Besredka and L. Gross, "De L'immunisation Contre le Sarcome de la Souris par la Voie Intracutanée," *Annales de l'Institut Pasteur, 55* (1935), pp. 491–500.
110. L. Gross, "The Vertical Epidemic of Mammary Carcinoma in Mice," *Surgery, Gynecology and Obstectrics, 88* (1949), pp. 295–308.
111. L. Gross, *Oncogenic Viruses*, 1st ed., (London: Academic Press, 1962).
112. L. Gross, "Susceptibility of Suckling-Infant, and Resistance of Adult Mice of the C3H and the C57 Lines to Inoculation with Ak Leukemia," *Cancer, 3* (1950), pp. 1073–1087; and L. Gross, "Pathogenic Properties and Vertical Transmission of the Mouse Leukemia Agent," *Proceedings of the Society for Experimental Biology and Medicine, 78* (1951), pp. 342–348.
113. M.B. Shimkin, "As Memory Serves. An Informal History of the National Cancer Institute," *Journal of the National Cancer Institute, 59* (1977), p. 559.
114. L.W. Law, T.B. Dunn, and P.J. Boyle, "Neoplasms in the C3H Strain and in F1 Hybrid Mice of Two Crosses Following Introduction of Extracts and Filtrates of Leukemic Tissues," *Journal of the National Cancer Institute, 16* (1955), pp. 495–519.
115. L. Gross, "Difference in Susceptibility to Ak Leukemic Agent Between Two Substrains of Mice of C3H Line," *Proceedings of the Society for Experimental Biology and Medicine, 88* (1955), pp. 64–66.
116. Gross, *Oncogenic Viruses.*
117. J. Furth, *Proceedings of the Society for Experimental Biology and Medicine, 93* (1956), pp. 165–172.
118. Law, Dunn, and Boyle, "Neoplasms in the C3H Strain," *Journal of the National Cancer Institute, 16* (1955), pp. 495–519.
119. L. Gross, "Development and Serial Cell-free Passage of a Highly Potent Strain of Mouse Leukemia Virus," *Proceedings of the Society for Experimental Biology and Medicine, 94* (1957), pp. 767–771.
120. E. Stewart, B.E. Eddy, and N. Borgese, "Neoplasms in Mice Inoculated with a Tumor Agent Carried in Tissue Culture," *Journal of the National Cancer Institute, 20* (1958), pp. 1223–1243.
121. J.B. Moloney, "Preliminary Studies on a Mouse Lymphoid Leukemia Virus Extracted from Sarcoma 37," *Proceedings of the American Association for Cancer Research, 3* (1959), p. 44.
122. R.A. Manaker, L.R. Sibal, and J.B. Moloney, "Scientific Activities at the National Cancer Institute: Virology," *Journal of the National Cancer Institute, 59* (1977), pp. 623–631.
123. L. Gross, *Oncogenic Viruses,* 2d ed., 1970.
124. National Cancer Institute, *Annual Report of Activities, Fiscal Year 1968.*
125. National Cancer Institute, *Annual Report of Activities, Fiscal Year 1973.*
126. N. Wade, "Scientists and the Press: Cancer Scare Story That Wasn't," *Science, 174* (1961), pp. 679–680.

THE AUTOMATED LABORATORY

The Generation and Replication of Work in Molecular Genetics

PETER KEATING
CAMILLE LIMOGES
Départment d'histoire, Université du Québec à Montréal,
Centre Interuniversitaire de Recherche sur la Science et la Technologie (CIRST),
Montreal, Canada

ALBERTO CAMBROSIO
Department of Social Studies of Medicine, McGill University, Montreal, Quebec

Introduction

Research in molecular genetics is undergoing a rapid and in-depth transformation, epitomized by the massive investment of money and equipment in the Human Genome Project (HGP). According to Walter Gilbert, the nature of this transformation can be expressed as follows: "The benefits of the Human Genome Project are the benefits that will accrue from *organization and scale.*"[1] At the heart of this new organization and scale lies new instrumentation and, in particular, automated instrumentation.

Automation in the biomedical field is not restricted to genetics, nor are the tasks targeted for automation merely the routine. They range from the simplest clinical assay, such as the Pap smear, which even "young girls" can do, to those which practically nobody can do well. In the latter category, one might include, for example, "[T]the blackest of biological black arts," the growth of macromolecular crystals, which, it seems, has also been slated for automation because of the hit-or-miss quality of its performance.[2] Indeed, molecular biology as a whole has been singled out as ripe for automation partly because of the lack of routine. As one projector recently put it:

Molecular biology techniques are very green-fingered and slow, involving a lot of "the expert's" time. Often, the only way to learn a technique is to go to someone's lab and learn it – it's often even difficult to get it from a [published] protocol. The idea [of automation] is to take the mystique out.[3]

125

Michael Fortun and Everett Mendelsohn (eds.), The Practices of Human Genetics, 125–142
©1999 *Kluwer Academic Publishers. Printed in Great Britain.*

Flushing out tacit knowledge, it seems, is at once the great advantage and the great need for automation in the sciences.

Accordingly, the language that scientists use often sounds as if it were drawn from the sociology of work. Walter Gilbert, for instance, in an attempt to determine what ought to count as science in the HGP, points out that "if large-scale sequencing is to work, it must be treated not as science but as a *production job*," a "pure technological problem quite apart from interpreting sequence." The problem of producing a large amount of sequence is "like building an automobile," and that work ought to be done by "*production workers*.... It is not done by research scientists."[4]

In a similar fashion, Leroy Hood tells us that "for sequencing the human genome, developing the technologies requires science and engineering. Obtaining the complete sequence is *a production-line effort,* and using the resultant data is science."[5] In Hood's view, scientific research is becoming the activity of a new form of laboratory, a *collective of human and nonhumans, inextricably composed of scientists, machines and technicians.* Such a move implies a change of emphasis from the research team led by a single researcher to a new type of center, epitomized by Hood's own "Center for Molecular Biotechnology," and a new type of biologist. Geared to the production of new tools and their application to leading-edge biology, the new biologist is envisioned as a sort of collective cyborg, which would be interdisciplinary, vertically integrated, system integrated and (of course) computerized.

Indeed, according to Hood, biology's "most exciting frontiers will require expensive and complex new instrumentation" and will thereby lead biologists to "the realm of big science."[6] Moreover, the new instrumentation "requires an organizational structure quite distinct from those commonly found in academia: a "sequencing production line must be established," "separate groups within the sequencing project should be established to govern quality control, troubleshooting, sequence closure and assembly, and protocol development and testing," and computational problems in large-scale sequencing will demand the establishment of "laboratory management systems."[7] In this view, automated sequencers and synthesizers are tools,[8] but tools that change – in depth – the way research is conducted and how actors put their act together.

The promoters of the new type of laboratory organization may have found their most enthusiastic advocate in Daniel Cohen, the head of the French Généthon, part of the Centre d'Étude sur le Polymorphisme Humain [CEPH]. According to Cohen, "the CEPH is responsible for the introduction of 'productique' [production automation] into genetic research,"[9] heralding a new "time of gene factories," and the advent of "robot biologists," rendering the "stage of craftsman laboratory work" obsolete.[10] To make sense of these views of what laboratory automation – exemplified here by the automation of DNA sequencing – entails, we shall start by outlining some of the traits of this ongoing process, then move to consider the conditions and consequences of this transformation.

Laboratory Automation and the DNA Sequencer Technology

To gain some idea of the development of the field of laboratory automation in the last 20 years, consider the 1976 bibliography *Rapid Methods and Automation in Microbiology and Immunology.*[11] Based on a search through the previous eight year's literature, the bibliography followed the Second International Symposium on Rapid Methods and Automation, which had been spurred in part by recent screening demands for hepatitis B and rubella in pregnant women. Of the 2,978 items collected, a mere 182 fell into the authors' own category of automation. Moreover, most of these were ancillary devices for so-called automated culture systems. In other words, as only 6% of the items concerned automation, the inclusion of automation in the title of the bibliography was more projective than indicative.[12]

However, the 1980s saw a shift in rhetoric from automation *in* the laboratory to automation *of* the laboratory with the appearance of books such as *The Electronic Laboratory*[13] and periodicals such as *Advances in Laboratory Automation,* the organization of conferences such as "The Role of Robotics and Automation in Decoding the Human Genome"[14] and the institution of such programs as the European Eureka Labimap 2000 Program to automate techniques in molecular biology.[15] The FDA acknowledged the pervasiveness of this trend by establishing a new set of guidelines called Good Automated Laboratory Practices (GALP).[16] Indeed, by the beginning of the 1990s, a representative of the American Laboratory Automation Standards Foundation (designed to serve the needs of the "laboratory automation community") went so far as to declare that "Laboratory automation is a bridging discipline that touches many other fields."[17] In other words, from a discourse on scientific activity, laboratory automation, it is claimed, has become an autonomous activity.

It was within this context that Leroy Hood and associates at Cal Tech formulated a strategy, at the end of the 1970s, to automate significant sections of the molecular biology laboratory, planning the production of a DNA sequencer, a DNA synthesizer and a protein sequencer and synthesizer. Let us restrict our remarks, for the present, to the DNA sequencer.

Research began on the instrument in 1978, shortly after the invention of the techniques presently used for sequencing. In an initial attempt, an optical/digital system was sought to replace human visual scrutiny of the DNA fragments layered in the acrilamide gel. Failure ensued, following several years of intensive effort. According the Hood:

> We probably spent the first three or four years in an approach to automation that turned out to be completely non-productive. And it turned out to be non-productive for a very interesting reason, and that is we attempted to do what many attempt to do in automating the procedure: take the chemistry as it was and just make a machine to do it. It turned out that the chemistry, as it was, wasn't appropriate for automation.[18]

In other words, automation was not achieved through the mechanization of the original protocol but in its reconceptualization. In this case, the reconceptualization resulted in the use of fluorescence tagging to provide a nonhuman form of recognition. It follows, than, that it is not always clear what it means to automate something. In other words, we may ask: what exactly *is* automated in DNA sequencing?

What Is the Object of Automation?

One possible answer has been proposed by the sociology of work, in which automation has been a central topic since at least Georges Friedmann's work in the 1940s,[19] and Harry Braverman's very popular study in the 1970s.[20]. In Braverman's work especially, the introduction of new technologies into the labor process has been represented as both a tragedy, insofar as it obviously "dehumanized" work, and a crime, insofar as it constituted a form of institutionalized theft, predicated on a process of deskilling wherein labor and laborers were divided until their skills could be incorporated into the physical design of, and their actions mimicked by, nonhuman devices. What has made this vision even more compelling is, in part, its more positive counterpart, the robot. As an image that pervades science fiction and magazines popularizing automated manufacturing, the robot, for anthropomorphic reasons, is considered the highest form of automation. More recently, within the sociology of scientific knowledge, debates over the limits of computerization have raised similar issues regarding the object of automation. In particular, Harry Collins and Hubert Dreyfus have recently presented opposing views about the criteria that make possible the computerization or automation of a given area of human activity such as scientific research.

According to Collins, automation is simply the ability of a machine to substitute for human action. That substitution is not an identical substitution but a functional substitution based on mimicry. A successful substitution is dependent on not only a social consensus as to what counts as the "same thing" or action but also on the ability of humans to make good the deficiencies that inevitably result from mimicry. It is in this sense that Collins claims that when machines "replace" humans, they do so only in the special sense that (a) they do not often do the same thing the humans did; and (b) they often appear to do so only because humans make good the deficiencies in the performance; and also that (c) when the replacement appears to work, it appears so because the human performance is itself modeled on or mimics a mechanical or machine-like performance.[21]

Whereas Collins believes that "replacement" is preceded by the constitution of a form of life involving machine-like behavior, Dreyfus believes that this disciplinarization is itself predicated on the possibility of regimenting an activity.[22] This possibility is not equally present in all realms of human endeavor or activity. It depends on the internal structure of the activity and is independent of the will or organization of the actors involved. It should not be concluded that Dreyfus is claiming to know in advance which domains of knowledge or activity can be automated and which cannot. Dreyfus's characterization of a field of activity is entirely pragmatic (nonessentialist) and thus does not lend itself to the extraction of internal criteria. Indeed, it may be said that, for Dreyfus, the description of a particular human activity for the purposes of automation, requires an endless and ultimately impossible task of explicitation. For example, Dreyfus claims that the activity "driving home from work," although a routine accomplishment, cannot be regimented to the extent that it can be accomplished semiconsciously and thus according to explicit rules or procedures. On this basis, Dreyfus claims that "driving home from work" cannot be imitated by a computer or automated.

Now Dreyfus says nothing about the "internal structure" of the activity of "driving home." He merely concludes *that* expert drivers are able to carry out the task with little conscious effort. It may thus be argued that Dreyfus is saying little more than, *until now*, computers have not imitated "driving home." He does not, therefore, exclude the possibility that if we change the process of driving home (as in the case of *Aramis*, as discussed by Latour),[23] it will be possible to make the activity digitizable. Dreyfus might argue that if we change the process, then we change the activity, and that therefore we have failed to do what we originally set out to do. Such a tactic would entail a theory of invariant descriptions, independent of social consensus, which would make such a judgment possible. However, Dreyfus presents no ontology of activity nor theory of description that would enable him to claim that, differently described, we no longer have the same activity and thus have failed to automate. In this sense, Dreyfus's argument is not about automation; it is about the impossibility of inventing an adequate description of human action. In this case, we would argue, Collins is partly right; we do automate certain activities even in the absence of a complete or exhaustive description. A reasonable facsimile is often acceptable. However, Collins's view itself is not without difficulties, for, to return to our original question, it is not exactly clear what is being automated.

Notice that, according to Collins's description of automation, machines always contain a residual deficiency. In this sense, like Dreyfus, Collins's analysis is based on a comparison between human action and machine action. The principle that machines are always less than human, and are somehow designed to replace human actions, is underscored, in Collins's case, by the contention that we are dealing with a "social prosthesis." Collins says, for example, "Computers have to be thought of as social prostheses – replacements for humans in com-munity."[24] Moreover, just as with a human prosthesis, the object of the social machine is to mimic prior human actions and thus constitutes an attempt to return to a prior state of affairs. But as prosthesis or imitation, machines and devices are not the "real thing." Hence "humans com-pensate for the deficiencies of artifacts in such a way that the social group continues to function as before."[25]

If we replace the term "social group" with "body," it becomes readily apparent that Collins has resurrected a form of thought that implicitly refuses to recognize a division of labor, even though it is a form of thought that is conscious of divisions among workers. In this sense, Collins's "social prosthesis" is reminiscent of early Greek theorizing, on two counts. First, it postulates machine performance deficiencies to be compensated by humans, the way Aristotle stressed the irremediable ontological inferiority of *techné* to *physis*. Second, it recalls the concept of the "Mighty Five" (machines), which represented tools as projections of the human body, and which conceptualized their workings as imitations of particular human actions, speech being the most basic.[26] And yet, Collins himself is clearly aware that machines are not really designed to replace parts of the human body or even "human beings in community"; his notion of mimicry shows merely that machines are capable or imitating humans at work. This is a considerably less ambitious intention for automation than that given under the previous description of social prosthesis.

Humans at work perform work routines, and these, *per se*, are clearly not reducible to something like human action or behavior, as they do not take place in the absence of tools. Moreover, work routines take place within a division of labor and are the scene of more than humans acting out fantasies of work. Work and its routines entail a variety of interactions between the living and the dead and the human and the nonhuman. What is replaced through automation or by a machine is not some fragment of the human body or some indescribable physical or mental gesture, but a way of *making* something; in other words, the object of imitation is not a way of doing but some means of producing something. In this sense, if we are to offer a description of how things get done in science, there will necessarily be a narrative continuity between machine action and human action. That such a story is anathema to Collins is clear, for he says, "Whether we come to speak of machines thinking without fear of contradiction will have something to do with whether my argument is more or less convincing than the argument of those who think of social life as continuous with the world of things."[27] We would argue, and this is not the contrary, that a description of scientific practices that establishes discontinuities between social life and "things" makes as little sense as saying that machines think.

Mimicry as Illusion

Automation need not mimic human operations, and, in fact, as we saw above in the case of the development of the gene sequencer by Leroy Hood's team, it may be more successful when it does not. Particularly telling in this regard is the comparison between differing approaches to automated gene sequencing. We are told, for example, that the "Americans" and the "Japanese" have followed opposing strategies. The "American approach," as we have seen, favors the redesign of experimental protocols to adapt them to existing strategies of automation. The "Japanese philosophy" has been described by Leroy Hood as follows:

They have chosen ... to take *existing techniques,* and to automate them directly by applying the principles of assembly-line production that have proven so successful in heavy industry. They are developing essentially a sequencing factory, in which each stage of the process is automated, and the operation of the instruments may be performed by relatively unskilled workers.[28]

Although we do not know exactly why, it is by now a well-known fact that "the Japanese" "grossly overestimated" the possibilities inherent in such a strategy.[29] We suspect there were difficulties similar to those encountered in the first phase of development of Hood's sequencer project: despite initial attempts to mechanize an existing protocol, it is ultimately not that protocol that was automated but, rather, a reorganized version. As Daniel Cohen was to prove with Généthon – we shall return to this later – the "Japanese" aim of developing a "sequencing factory" was not, as such, doomed from the outset. But success seems to have been predicated on creating machines that fit into an entirely new division of labor and not in attempting to replicate existing laboratory tasks.

Successful automation is not, therefore, an automated mimicking of human operations by a more efficient machine, but a *substitution* of one process (involving more than humans) by another. What we have is not a deskilling of humans by embedding human skills in an automaton, but the creation of an emerging new field of operations that redistributes actions between humans and machines and between the humans themselves. The emphasis here is on actions, not skills. As Stephen Barley has pointed out, studies that have followed Braverman and focused on skills have adopted a "narrow if not misleading" viewpoint with the result that they inevitably distract one from the potential of social action.[30] To deskill a task is not necessarily to deskill a worker, and empirical studies have shown repeatedly that, far from being derivative of the actuation of preexisting skills, occupational dynamics are strongly context-bound.[31]

If the object of automation is the production of results in a new way, further questions may be asked: What is the identity of these results? Are those obtained with a DNA sequencer the same as those previously obtained by manual sequencing? The answer seems to be yes. In both cases, we have new information on the sequence of some bases. On the other hand, could this be mere appearance? Could one set of results merely mimic the other?

Although sequences read on an autoradiogram taken on an electro-phoretic gel look strikingly similar to the columns and rows on the screen of a microcomputer – and much work has been done to ensure that similarity – we also know that those two end results have not been produced in the same way at all (see Fig. 1). Computer imaging comes at the end of a translation, through digitization, which has no equivalent in manual sequencing. Indeed, the type of representation adopted for the computer screen might have been entirely different, since we know that other automated machines produce chromatograms; these are more difficult to read than autoradiograms, however, because biologists are unfamiliar with them.[32] Chromatograms can be "read" or, rather, interpreted, as giving the "same" results as autoradiograms, but that merely means that the "results," the information on the order of bases, is the product of an interpretation, a translation, by the "reader." In fact, the same situation obtains when one compares "data" on an autoradiogram or on a computer screen; the inscriptions look identical, but they are not the same. One just mimics the other.

Once again, analogies with the conventional representation of automation in manufacturing are misleading because that representation is wrong. Automated manufacturing processes do not reproduce human operations; even robots do not do that beyond crude anthropomorphic similarities. Moreover, the claim that automated industrial processes produce the same merchandise as conventional production is inexact. A car built by robots may look and seem to function just like a car put together by humans using tools, but that does not mean that we still build cars as we did before robots. Anyone who has attempted to repair a car in the last ten years knows that.

Automated sequencers are interesting precisely because they produce new results and not just more of the same. An automated sequencer produces inscriptions, which are valued because of their originality, and although they can be interpreted the same way as marks on an autoradiogram, they are not really the same. Since one of the advantages of automated sequencing is "improved accuracy," automated sequencing data cannot be both the same as manual sequencing data and more accurate.

From Sequencers to the Automated Laboratory, or Is the HGP Science or Technology?

Automated sequencers do more than manual sequencers. Those developing robotics equipment for the HGP, for example, claim that:

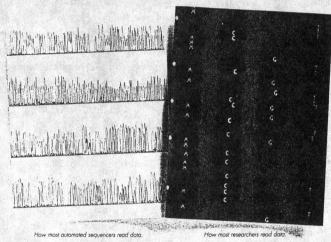

Wouldn't It Be Nice If Automated DNA Sequencers Spoke The Same Language As Researchers?

How most automated sequencers read data.   *How most researchers read data.*

Why be forced into reading chromatogram-like peaks when you can see your data the way you're used to seeing it: as sequence ladders.

The new BaseStation™ Automated DNA Sequencer from Millipore reads, analyzes and presents data as familiar four-lane sequence ladders. In fact, it's the only sequencer to display data this way.

The system is able to accomplish this because of the novel way it looks at DNA fragments in a gel. Light emitted by fluorescing DNA is captured inside a charge coupled device (CCD) camera. The CCD is a matrix array of thousands of individual detectors, each capable of registering a single photon of light. This lets the camera see DNA as well-defined bands, allowing for accurate interpretation of data.

Up to 20 kb of primary sequence can be analyzed every day. Over 500 bases are run per sample, with 98% of the first 400 identified

unambiguously by the BaseStation system.

The BaseStation system also analyzes your data in real time. A Fluorogram™ (similar to an autoradiogram) is displayed on screen as the bands move down the gel, so you can assess the quality of the sequencing chemistry. A typical run takes 7 hours, including pre-electrophoresis and loading. When it's done, so is your analysis.

And our DNA Sequence Manager makes reviewing data simple. It automatically generates "contigs" from your sequence sets. And to aid in resolving areas of ambiguity, the system can also display the original Fluorogram image.

For a demonstration or more complete information, call 800-225-1380 in the U.S.; in Japan call (81) (3) 3474-9111; and in Europe call (33)1.30.12.72.34.

MILLIPORE

© 1992 Millipore Corporation

Circle No. 12 on Readers' Service Card

Fig. 1 Mimicking autoradiograms. In this 1992 advertisement for an automated DNA sequencing machine, the product's advantages are promoted by contrasting its data representation technique (top right) with the data displays featured by competing machines (top left). The advertised instrument imitates the visual results of manual sequencing. (Reproduced with the kind permission of Millipore Corporation; original is in color)

Integrating robotics technology with advanced instrumentation will facilitate an enormous increase in accuracy and provide the opportunity for skilled researchers to concentrate on experimentation rather than the tedious operations that can, in fact, be done better and more efficiently by machines.[33]

Gilbert and Hood have also made this point. In fact, more than just chasing routine from the laboratory, automated sequencing is seen as making certain kinds of research and thus certain kinds of routines possible.[34] Moreover, sequence data must not be treated as "atomic" facts but, rather, be seen as part of a interconnected system of data production. Changing an element in this network changes the network and, hence, the sense accorded the items in the system.[35] Nor should the "increase in accuracy" be understood in a narrow, technical sense. The machines introduced in the nineteenth century embodied the ideal of the indefatigable observer and the "promise of images uncontaminated by interpretation."[36] Machines today, in the late twentieth century, offer unprecedented possibilities of "tampering" with data and thus transforming the value attributed to scientific images.[37] In this way, the introduction of machines has contributed to a subtle but nonetheless decisive shift in the epistemic underpinnings of scientific practice.

In a similar fashion, the introduction of gene sequencers called for the automation of the entire chain of protocols, from DNA extraction and purification machines to sequence analyzers and chromosome sequence cartographers. This is what Daniel Cohen and the Bertin company undertook at Généthon, when they constructed their "chaînes modulables de robots" (flexible robot production lines).[38] The development of a whole system of machines, reconfiguring the entire "production" process into an automated laboratory, or "gene factory," follows dynamics similar to what happened in the history of technology or during the industrial revolution. The mechanization of any step in a production process creates strains on the others;[39] "the different operations of a given industry are like a set of solidary movements, submitted to the same rhythm."[40] The gene sequencer is not simply an additional tool to be applied to routine practices but the basis for a complete reconceptualization of laboratory work in molecular genetics and the fulcrum of its present reorganization.

On a more general level, automation of the laboratory is presented as making possible a different kind of science, one which embraces the repetitive as an initial stage of inquiry as opposed to a tedious process of working through. It is argued that:

These devices [for automation] allow the scientist to perform more aggressive experimental protocols that involve either large numbers of samples or unique compounds requiring lengthy preparations.[41]

In other words, the very existence of technologies of automation is seen to generate approaches to problems whose solution had heretofore been considered impossible.[42]

Given the foregoing, one might wonder to what extent the HGP can be considered science rather than technology. According to one of the architects of the project, "sequencing the human genome is not actually science because it is not an endeavor concerned with posing questions and formulating experiments to answer them."[43] Nonetheless, the HGP remains a research program conducted by scientists and, as such, is also portrayed as science. According to one defender of the initiative, the problem lies in the fact that "people are uncomfortable with research projects designed to improve technology as opposed to a research project designed to extract a few facts."[44]

A discussion of automation in the sciences may seem somewhat paradoxical for those who believe in the linear model of science leading to technology, as, in this view, science itself is the ultimate source of automation. It is true that the linear model has been discredited, and there are those who now believe that the development of technology is largely independent of the development of science. However, there is also considerable opinion within the sociology of science that sees little difference between science and technology, and that, consequently, views the activities that produce them as fundamentally the same. In that line of thinking, the paradox reemerges in revised form. According to Star, for example, "Science and work are the same activity, but differently reported."[45] Similarly, Shapin has recently proposed that "the laboratory is now to be understood more on the model of a workshop and scientists are now to be understood more on the model of technicians."[46] If these suggestions are to be taken literally – and we suggest they should not – then we would find that the transformation of the workplace as described by labor process analysts applies equally well to the laboratory.

Indeed, if there is automation *in* the sciences, it should mean that we are dealing with much more than just the automation of those tasks that have become so routinized as to be little more than mechanical undertakings best pursued by machines. It is true that the repetitious and mechanical nature of much of modern biomedical research is a recurrent theme among promoters of automation. According to them, there are a number of "highly skilled people" presently involved in "extremely tedious, repetitive and expensive procedures" – not good news, perhaps, for people planning a career in science. Whether or not this is the norm, given the "creative" nature of the scientific enterprise, those who are involved in the sales of automation devices would do well to invoke the standard reasons for purchasing their equipment, that is, increased productivity, or what is more diplomatically termed "throughput." Moreover, it is also natural that equipment producers would wish to see their devices as more than just peripheral to the conduct of science. Therefore, promotional phrases such as the "mind-numbing demands of modern day-to-day biology"[47] or the "drudgery" of molecular biology, whose "tedious" techniques require "boring," "mental gymnastics"[48] are used to explain the need and demand for automation. From the laboratory itself, consider the following description of life before and after automatic sequencing:

> We spent about a year doing manual sequencing, and then we were not very impressed with the technology at all.... It was clear that manual sequencing techniques were not going to get us very far ... People in the lab who have since switched to automatic sequencing were threatening to leave if they had to keep doing that much longer. It's tedious, not very rewarding work.[49]

However close these complaints may bring us to the shop floor, there are needs that are peculiar to science and specify the process within this sphere of activity. Consider, for instance, the following quote: "The primary objective in the use of automation is standardization, stimulated by the need for repetitive, highly accurate determinations."[50] We see here that neither cost nor output are important and that the repetitive nature of scientific tasks is generated *from within science*. However, there is more, for automation can be seen as a process of standardization that provides scientists themselves with an insight into the nature of their work. The previous quote continues: "Through standardization, it is possible to determine the critical techniques required for excellent laboratory results and then to incorporate these techniques permanently to provide highly consistent results."[51]

While it is generally held that industry seeks the production of the same in the form of a standardized commodity, science, its devices, and its experimental systems seek the production of the new and the different. Ironically, routinization is clearly seen by the "automation community" as one of the privileged means toward this end. Through automation, heretofore unknown or invisible "critical techniques" are uncovered, making possible the achievement of the primary objective of standardization. Automation thus results in both the routinization of a science and the means through which differences are created.

Conclusion

The automation of the laboratory, as exemplified in the automated sequencer, an inscription device that is part of a complete reconfiguration of laboratory work, does not fall within the traditional understanding of the relationship of automation with the division of labor. What we have is not a progressive division and subdivision of labor so that preexisting tasks are isolated and simplified in a way that makes them more amenable to machine function. As we have pointed out, we do not have an incorporation of atomized human skills into a machine, but a complete substitution of one process by another. This is so much the case that it is questionable, as we have also seen, to what extent the end results of the two processes – human versus machine performance of protocols – are the same, even when they are customized to mimic one another. They are at least different enough to engender not only a more rapid and lower-cost activity but also changes in the content of science and in scientific approaches. Indeed, Daniel Cohen speaks of a "new race of researchers."

In this respect, to reduce the significance of automation to a deskilling process is to lose sight of its most striking consequences: it redefines the boundaries between humans and nonhumans; it redistributes humans in the laboratory and redefines their skills and functions. For example, although the human eye looks at inscriptions on the computer screen that mimic those on an autoradiogram, the work done and that which remains to be done is no longer the same. The digitization "behind" the computer imaging also makes it possible to manipulate data in computerized form. Data then become immediately transportable, comparison between sequences immediately automatizable, and so on.

Finally, automation redistributes and reconfigures what is epistemic and what is technological. As we have seen, labeling genome sequencing as science or as technology is a debate that is very much alive among scientists themselves. Inside the experimental system, it would seem, the epistemic (the sequence, which is new knowledge) is considered by some to be technological in the sense that it can no longer be the object of a Ph.D. dissertation.[52] What was once deemed research, becomes a technological process; the handling of data, no more a "research" topic, though an integral part of research process. This is why Hood has claimed that discoveries are no longer made by individuals, but as big science, in a collective of scientists, machines, and technicians. Technology is not excluded but taken for granted, as an indispensable ingredient of the process. Similar sentiments have been expressed by

Julian Davies of the Institut Pasteur, according to whom the techniques determine the science agenda. Thus, "when asked how to solve a problem in biology, ... a graduate student at the end of the 1970s would say, 'clone it'; ...in 2001, [his response probably would] be, 'send it to the service center.'"[53]

Acknowledgments

Research for this paper was made possible by the Social Sciences and Humanities Research Council of Canada, grant 410-91-1935. Earlier versions of this paper were presented at the Joint Conference of the Society for Social Studies of Science and the European Association for the Study of Science and Technology (Göteborg, Sweden, August 12–15, 1992) and at the Sociology of the Sciences Yearbook Conference "The Practices of Human Genetics: International and Interdisciplinary Perspectives," held in conjunction with the International Society for the History, Philosophy, and Social Studies of Biology Conference (Brandeis University, Waltham, Massachusetts, July 14 –15, 1993).

Notes and References

1. Walter Gilbert, "Human Genome Sequencing," in *Biotechnology and the Human Genome,* ed. Avril D. Woodhead and Benjamin J. Bernhart (New York, Plenum Press, 1988), p. 35 (emphasis added).
2. Bio/Technology Staff Report, "The Laboratory Robot," *Bio/Technology, 5* (1987), p. 463.
3. John Hodgson, "Molecular Biology in 2001," *Bio/Technology, 8* (1990), p. 190.
4. Walter Gilbert, quoted in Leslie Roberts, "Large-Scale Sequencing Trials Begin," *Science, 250* (1990), p. 1336 (emphasis added).
5. Lloyd Smith and Leroy Hood, "Mapping and Sequencing the Human Genome: How to Proceed," *Bio/Technology, 5* (1987), p. 934 (emphasis added).
6. Smith and Hood, "Mapping and Sequencing," p. 939.
7. Tom Hunkapiller, R. J. Kaiser, B. D. Koop, and Leroy Hood, "Large-Scale and Automated DNA Sequence Determination," *Science 254* (1991), p. 66.
8. Melissa Hendricks, "The Big Biology of Leroy Hood," *Johns Hopkins Magazine,* February 1990, pp. 41– 50.
9. Daniel Cohen, *Les Gènes de l'Espoir. A la Découverte du Génome Humain* (Paris: Robert Laffont, 1993), p. 100.
10. *Ibid.,* p. 99.
11. Wendy J. Palmer and Suzanne E. Le Quesne, eds., *Rapid Methods and Automation in Microbiology and Immunology,* (London and Washington: Information Retrieval, 1976).

12. But is it not possible that such will always be the case with automation, in the sense that every task that is automated will eventually become an instrument, which in turn will require automation? Or is it rather the case that automation itself is not the end of the process but, as already indicated, merely an intermediary in the pursuit of dedicated instruments?

13. Raymond E. Dessy, *The Electronic Laboratory. Tutorials and Case Histories in Laboratory Automation* (Washington, D.C.: American Chemical Society, 1985).

14. Dan W. Knobeloch, Carl E. Hildebrand, Robert K. Moyzis, Jonathan L. Longmire, Karl M. Sirotkin, and Tony J. Beugelsdijk, "Robotics in the Human Genome Project," *Bio/Technology, 5* (1987), pp. 1284–1287.

15. The companies and centers involved in this project are Amersham International, Imperial Cancer Research Fund, Centre d'Études sur le Polymorphisme Humain, and Bertin. Reported in Hodgson, "Molecular Biology in 2001," p. 190.

16. Christopher Anderson, "Easy-to-Alter Digital Images Raise Fears of Tampering," *Science, 263* (1994), pp. 317–318.

17. R. Lysakowski, "Toward Laboratory Automation System Standards," *American Biotechnology Laboratory,* February 1992, p. 42.

18. Leroy Hood, interview by Ramunus Kondratas, October 19, 1988. Smithsonian Videohistory Program, Smithsonian Archives, Washington, D.C. p. 11. We thank Ray Kondratas for facilitating our access to the transcripts of these interviews.

19. Georges Friedmann, *Problèmes Humains du Machinisme Industriel* (Paris: Gallimard, 1946).

20. Harry Braverman, *Labor and Monopoly Capital: The Degradation of Work in the Twentieth Century* (New York and London: Monthly Review Press, 1974).

21. Harry M. Collins, *Artificial Experts: Social Knowledge and Intelligent Machines* (Cambridge, MA: MIT Press, 1990), p. 9.

22. See Hubert L. Dreyfus, "Response to Collins, *Artificial Experts,*" *Social Studies of Science, 22* (1992), pp. 717–726; and Harry Collins, "Hubert L. Dreyfus, Forms of Life, and a Simple Test for Machine Intelligence," in *Ibid.,* pp. 726–739.

23. Bruno Latour *Aramis ou l'Amour des Techniques* (Paris: La Découverte, 1992).

24. Collins, *Artificial Experts,* p. 216.

25. *Ibid.*

26. Jean-Pierre Vernant, "Remarques sur les Formes et les Limites de la Pensée Technique chez les Grecs," in *Mythe et Pensée chez les Grecs* by Jean-Pierre Vernant (Paris: Maspéro, 1965), pp. 227–247.

27. Collins, *Artificial Experts,* p. 224.

28. Smith and Hood, "Mapping and Sequencing," p. 936 (emphasis added).

29. See Daniel J. Kevles, "The Historical Politics of the Human Genome," in *The Code of Codes,* ed. Daniel J. Kevles and Leroy Hood (Cambridge, MA: Harvard University Press, 1992), p. 36; see also Tracy L. Friedman, "The Science and Politics of the Human Genome Project," Senior Thesis, Woodrow Wilson School of Government, Princeton University, April 1990, pp. 50 – 51.

30. Stephen R. Barley, "Technology, Power, and the Social Organization of Work: Towards a Pragmatic Theory of Skilling and Deskilling," *Research in the Sociology of Organizations, 6* (1988), p. 53.

31. *Ibid.*

32. Thierry Damerval, "ADN: l'Ère des Séquenceurs Automatiques: Propos Recueillis auprès de Dick Barker," *Biofutur, 118,* December 1992, p. 20.

33. Knobeloch et al., "Robotics in the Human Genome Project," p. 284.

34. Curiously, these projects are often made to appear mundane; composed of individual tasks that are equally as "dull," such as, for example, the enzyme-linked immuno-absorbent assay that a robotic expert, with a view toward automation, has analyzed into 1,200 separate steps. "I know because I sat down one day and counted them," stated Lawrence Haff of Perkin-Elmer's laboratory robotics division in "The Laboratory Robot," *Bio/Technology Staff Report,* p. 459.
35. Michel Callon, "Is Science a Public Good?" *Science, Technology, and Human Values,* Science, Technology and Human Values, 19 (1994), pp. 385–424.
36. Lorraine Daston and Peter Galison, "The Image of Objectivity," *Representations, 40* (1992), pp. 119 –120.
37. Anderson, "Easy-to-Alter;" see also William J. Mitchell, *The Reconfigured Eye: Visual Truth in the Post-Photographic Era* (Cambridge, MA: MIT Press, 1992).
38. Daniel Cohen, *Les Gènes de l'Espoir,* p. 100.
39. David Landes, *The Prometheus Unbound* (Cambridge: Cambridge University Press, 1970), p. 81.
40. Paul Mantoux, *La Révolution Industrielle au XVIIIe Siècle* (Paris: Éditions Genin, 1959), p. 205.
41. Knobeloch et al., "Robotics in the Human Genome Project," p. 1284.
42. The simplest and yet the most recent form of automation is robotics, a mechanical device that mimics hand movements. First produced in the beginning of the 1980s, following market research on production bottlenecks, robots made a number of inroads into both the chemical and the biological laboratory in the area of sample preparation. Borrowing a 1920s concept from chemical engineering, the industrial leader trademarked what it termed "Laboratory Unit Operations," all ten of which can presently be roboticised. They are: Weighing, Grinding, Liquid Handling (Pipetting, Dispensing, Diluting), Conditioning (Heating/Cooling, Mixing), Manipulation, Measurement, Separation, Control (Decision making), Data reduction, Documentation; see Mary Jean Pramik, "Laboratory Robots in the Forefront of Automation Trend in Biotechnology," *Genetic Engineering News,* September 1988, p. 7. From what we already have seen, the status of the robot is not exactly clear in the literature. First of all, does a robot simply replace deskilled labor, or does it make impossible tasks possible by virtue of inhuman repetition? Second, even though, at present, robots are considered intermediaries in a process that ultimately should lead to an automated device for the task at hand, their presence in the laboratory is considered by some as an incentive to redesign laboratory protocols so as to maximise their use. Presumably, this would entail the design of experiments with a maximum number of dull and repetitive steps. (*Ibid.,* p. 463.)
43. Smith and Hood, "Mapping and Sequencing," p. 934.
44. Stephen S. Hall, "How Technique is Changing Science," *Science, 257* (1992), p. 346.
45. Susan Leigh Star, *Regions of the Mind* (Stanford: Stanford University Press, 1989), p. 198.
46. Steve Shapin, "The Invisible Technician," *American Scientist, 77* (1989), p. 563.
47. Bio/Technology Staff Report, "The Laboratory Robot," p. 459.
48. Ricki Lewis, "DNA Software Takes the Drudgery out of Molecular Biology," *The Scientist,* September 16, 1991, p. 23.
49. J. Craig Venter, interview by Ramunus Kondratas, March 1990, Smithsonian Videohistory Program, Smithsonian Archives, Washington, D.C. p. 27.
50. Knobeloch et al., "Robotics in the Human Genome Project," p. 1284.
51. *Ibid.*

52. To put it in another way, "People continue to work in the old-fashioned way, but I have my doubts that it advances careers. It used to be that you could get a job if you could sequence DNA. Now, it you sequence too much, you probably cannot get a job *because you have done nothing interesting.*" James D. Watson, "A Personal View of the Project," in *The Code of Codes*, ed. Kevles and Hood, p. 170 (emphasis added).
53. Hodgson, "Molecular Biology in 2001," p. 192.

HANS NACHTSHEIM, A HUMAN GENETICIST
UNDER NATIONAL SOCIALISM,
AND THE QUESTION OF FREEDOM OF SCIENCE*

UTE DEICHMANN

Institut für Genetik, Universität zu Köln, Köln, Germany

Scientists in Germany after World War II faced a short period of international isolation. There was no official boycott comparable to the one that was implemented after World War I. But in many cases, scientific connections with Germans were cut, Germans were not invited to international congresses, or they were invited only individually, according to their assumed integrity during the National Socialist (NS) past. In particular, geneticists – above all, human geneticists – were criticised for having supported the ideological ends of Nazi Germany and for having provided the scientific basis for genocide programs. Hermann Muller used the term "prostitution of science"[1] to describe the readiness with which many geneticists in Germany used their field to support the Nazi ideology and murderous practice. To name only the leading human geneticists who were criticised in this respect: Eugen Fischer, professor of anthropology in Berlin, and who, until 1942, was director of the Kaiser Wilhelm Institute (KWI) for Anthropology, Human Genetics, and Eugenics; Otmar von Verschuer, professor of genetic care and race hygiene, who became Fischer's successor in the KWI in 1942; and Fritz Lenz, professor of race hygiene and department head of the KWI. All three men were known to have supported not only racial hygiene but also the anti-Jewish and racist views and laws of the regime. The extent of the close cooperation of von Verschuer with Joseph Mengele in Auschwitz became known only much later.

Did all human geneticists, or as they were called then, anthropologists, race hygienists, and psychiatrists contribute to anti-Semitism, racism,

---

*This paper is based on my larger study *Biologists Under Hitler* (Cambridge, MA: Harvard University Press, 1996).

Michael Fortun and Everett Mendelsohn (eds.), The Practices of Human Genetics, 143–153
©1999 *Kluwer Academic Publishers. Printed in Great Britain.*

and murder in the Third Reich? Hans Nachtsheim is reputed to be a prominent exception. An internationally renowned animal geneticist, he had turned to research in the genetic diseases of animals and men. He was not a member of the Nationalsozialistische Deutsche Arbeiter Partei, and geneticists in the United States and Great Britain were convinced that he was an anti-Nazi. According to Paul Weindling, for example, emigré geneticists such as Richard Goldschmidt and Hans Grüneberg vouched for his integrity.[2] Thus he earned the confidence of the Allies and of geneticists in Western countries.

There was no general invitation to Germans to take part in the international genetics congress in Sweden in 1948, only personal invitations to geneticists known not to have been Nazis. Nachtsheim was asked by Gert Bonnier, secretary general of the congress, to list the German geneticists who qualified. He named 18 geneticists, among them only two human geneticists, Gustav Becker and Friedrich Curtius, and they each received a personal invitation.[3] In Germany after 1945, Nachtsheim became the central figure in founding human genetics.

If we look at Nachtsheim's career and research during the National Socialist period, we find confirmation for the widespread belief that Nachtsheim was not one of the anti-Semitic or racist human geneticists of the Third Reich. The claim, however, that he behaved with integrity politically and personally and that National Socialist ideology did not enter into his research is not tenable. I want to shed some light in this paper on the practice of normal human genetic research in a society known to have abolished the value of equal human rights.

Hans Nachtsheim was born in 1890, studied zoology and received his Ph.D. at the Institute for Zoology of the University of Munich in 1913, and was *habilitated* in 1919.[5] In 1923, he became assistant and associate professor at the Institute for Genetics of the Agricultural College in Berlin (since 1935, part of the university), headed by Erwin Baur. In 1926, he went to the United States for one year on a fellowship from the Rockefeller Foundation, spending most of the time in Thomas H. Morgan's lab at Columbia University. As he was neither Jewish nor an outspoken left winger nor a liberal, his position was not endangered by the Law for the Restoration of the German Civil Service, implemented on April 7, 1933, shortly after the Nazis came to power. But for a while, Nachtsheim feared losing his position as an assistant at the Institute, perhaps as soon as 1934, because of plans to abolish long-lasting assistant positions. He contacted colleagues in the United States, therefore, asking about work and there was talk about giving him a position at the University of

Wisconsin with financial support of the Rockefeller Foundation.[6] When his position in Berlin was prolonged in October 1933, however, he decided not to go to Madison.[7] His position was continued again in 1937, due to a positive political review by the *Dozentenschaft* (the faculty body).[8] The opinion stated that Nachtsheim had always acted in favor of the national state – for example, he had hoisted the black-white-red flag (the flag of the German Reich, which indicated an anti-Weimar Republic attitude) during the Weimar period – and that he was an outstanding geneticist. In 1940, he was invited by Eugen Fischer to head a newly founded department for hereditary pathology at the KWI for Anthropology. Nachtsheim joined the Institute in 1941 and worked under Otmar von Verschuer, who became Fischer's successor in 1942, until 1945.

A close look at Nachtsheim's research reveals several distinct periods. In 1912, he began experimental work on sexual determination, leading later to work on the biology of the honey bee. Thus he was interested in problems of genetics very early on. When acting as a censor during World War I, he became acquainted with publications by T.H. Morgan. Then still during the war, he published a report about the work of Morgan's school. Later, he translated Morgan's book *The Physical Basis of Heredity* into German, and it was published in 1921. From 1921 on, he carried out genetic research on *Drosophila* and, after 1924, on small mammals as well. In 1933, he founded a new research discipline, comparative experimental genetic pathology, which involved experimental research on genetic diseases of small mammals. These studies were stated to be models for humans for situations in which experimental analysis was not possible. Nachtsheim considered this research to be important for human racial hygiene.[9]

Why did he change his research in 1933? According to Diane Paul, Nachtsheim wrote to Eugen Fischer on April 12, 1948 that the reason was the implementation of the "law for the prevention of genetically diseased offspring" on July 15, 1933.[10] According to this law, individuals could be sterilized if they suffered from a "genetic" illness, for example, feeble-mindedness, schizophrenia, or hereditary epilepsy. It is estimated that about 350,000 people were sterilized before 1939, most of them against their will. Working on the hypothesis that genetic diseases in humans and other mammals were caused by homologous mutations in the same gene, Nachtsheim looked for genetic diseases in small mammals, chiefly rabbits, and analysed their transmission and probability of manifestation. Among the diseases he investigated were Parkinson's, epilepsy, dwarfish growth, and Pelger anomaly.

Nachtsheim's research was closely related to the sterilization law. He was guided in his experiments by the idea that sterilization would prevent the spread of diseases that he considered dangerous from the race hygienist point of view. I don't want to criticize his attempts to improve the knowledge about genetic diseases. But in a society in which the carriers of genetic diseases were considered automatically to be inferior beings, Nachtsheim's research increased the already existing injustices based on genetic or assumed genetic differences. His findings fit the Nazi ideology of serving the *Volkskörper* and not the individual. For example, Nachtsheim called attention to the potential danger caused by the gene for Pelger anomaly, an irregularity in the form of the nucleus of white blood cells, which up to then had been regarded as harmless. He concluded that this gene was undesirable from the race hygiene point of view.[11] Regarding epilepsy, Nachtsheim tried to find methods to distinguish between the genetic and non-genetic forms of the condition. His aim was to develop a clear basis for the diagnosis of genetic diseases, which then should be "cured" by forced sterilization. He detected a recessive allele in rabbits that increased the incidence of epileptic seizure. He assumed that lack of oxygen in the brain was responsible for the seizure, which could be provoked by spasms in the arteries so he tested whether rabbits with hereditary epilepsy had seizures at a lower dose of Cardiazol (Metrazol), a drug which provokes spasms, than those with nongenetic epilepsy. Psychiatrists in the 1930s had already posed this hypothesis. Nachtsheim hoped that Cardiazol could thus serve as a means to diagnose hereditary epilepsy. But results were negative; there was no clear difference in the reaction to Cardiazol.

Nachtsheim could confirm his hypothesis that lack of oxygen was the primary cause for an epileptic seizure in epileptic rabbits, however, and he found that young and old animals reacted differently toward lack of oxygen.[12] In order to find out whether these differences could also be found in humans, Nachtsheim and his co-worker Gerhard Ruhenstroth-Bauer, a research fellow at the KWI for Biochemistry, tested whether 11- to 13-year-old children got epileptic seizures at low air pressures (corresponding to an altitude of 4,000 to 6,000 m).[13] They planned to continue these experiments with 5- to 6-year-old children in order to make the results comparable to those with rabbits. For these experiments, he used the air force's low pressure chamber. A preliminary report of these experiments was published in 1944,[14] but detailed publication never appeared, neither during nor after the war. Also, Nachtsheim published nothing else about epilepsy, at least until 1950.

Where did Nachtsheim and Ruhenstroth-Bauer get the children for the low-pressure experiments? Benno Müller-Hill, assuming that they might come from the euthanasia institution in Görden (Brandenburg), asked Ruhenstroth-Bauer about the experiments. Ruhenstroth-Bauer's lawyer replied stating that they were not experiments but clinical trials. They were permissible because the epileptic children came from an orphanage, and nobody had died or was physically harmed. Moreover, Ruhenstroth-Bauer claimed the results would be of therapeutic and prognostic value.[15]

This information was incorrect, as has become known recently. The human geneticist Gerhard Koch has published in his autobiography a letter that Nachtsheim wrote to him on September 20, 1942.[16] Nachtsheim stated that the children did come from Görden and went on to say: "...Mr. Ruhenstroth-Bauer had already contacted Wuhlgarten and was told that there are only grownup epileptics. He was referred to Görden, and from there we have received, due to the courtesy of *Obermedizinalrat* Dr. Brockhausen, six epileptic children (four genuine, two symptomatic epileptics) with whom we carried out experiments in the low pressure chamber of Prof. Strughold last Friday. These experiments had, however, the same negative results as the experiments with adult epileptics, conducted by Gremmler....We should test 5- to 6-year-old children in addition, but, for the time being, this is not possible, because there are no children of this age in Görden."[17]

The youth department of the Psychiatric *Landesanstalt* Görden near Berlin was headed by Prof. Hans Heinze, a referee for euthanasia. The children were, as in the youth departments of other psychiatric hospitals, selected for euthanasia, and murdered either in Görden itself or in Brandenburg.[18]

We do not know about the fate of the children who were experimented on by Nachtsheim and Ruhenstroth-Bauer. There is evidence of one low-pressure experiment, however, for which the date and the age of the child correspond to Nachtsheim's description. According to Dr. Hans-Hinrich Knaape,[20] a low-pressure experiment was carried out on an 11-year-old girl of this *Landesanstalt* in order to find out whether she suffered from hereditary or symptomatic epilepsy.[19] The report says that this investigation took place at the low-pressure chambers of the air force on September 17, 1943, and that the low pressure corresponded to an altitude of 4000 m. This same girl had survived by 1945. Nachtsheim made use of the fact that during National Socialism "inferior" people in psychiatric hospitals or in concentration camps were available as "material" for medical experimentation. He had to cooperate with physicians who carried out euthanasia. Neither he nor his colleagues ever mentioned this fact after 1945; only Gerhard Koch's publication of his letter in 1993 made Nachtsheim's connection to Görden evident.

A few points about Nachtsheim's postwar career and research – I mentioned already that he became a highly influential human geneticist in Germany; The Society for Anthropology and Human Genetics awards the Hans Nachtsheim prize for exceptional achievements in human genetics, one part of the prize for theoretical and the other part for practical research. Nachtsheim was one of the strongest advocates of eugenic sterilization in postwar Germany. After 1946, there was no legal basis for eugenic sterilization. Nachtsheim fought for the implementation of laws that would allow voluntary sterilization of genetically ill people as the strongest measure against the assumed dramatic increase of genetic diseases. In 1952, he even made it clear that he had fought for a legal basis for voluntary sterilization only for tactical reasons, and that he considered compulsory sterilization much more effective and sensible.[20]

I don't want to delve deeper into the eugenics issue and the continuity of an attitude, which in 1963 Alexander Mitscherlich described as "a monomaniacal concentration (*Fixierung*) of all eugenic practice…, which considers the active mutilating intervention of the sphere of the individual for the benefit of the state (*Allgemeinheit*) (be it defined as 'race,' 'nation,' or as the majority of the 'normals') as the last resort of truth."[21] For Mitscherlich and many others, there is a direct connection between the Nazis' sterilization law and their extermination of millions of people, both having originated in eugenic thinking based on "*Ausmerzen*" (eradication) as essential measures. I would rather look at Nachtsheim's research during the Nazi era from another perspective, that of scientific freedom.

Like many other West European and American geneticists, Nachtsheim fought against Lysenkoism in the USSR during the 1950s. According to him, communist totalitarianism led to the same suppression of the sciences that National Socialism yielded. Thus he viewed Lysenko's ideas about genetics to belong in the same "domain of pseudoscience as the race theories of the National Socialists."[23] He stated, "A totalitarian system cannot coexist with freedom of science. We have to continue our fight."[23] In order to protest against political restrictions and the lack of freedom in his work in the German Democratic Republic, he even left his position as professor of genetics at Humboldt University, which he had held since 1946. He accepted an offer to become professor of genetics at the Freie University in West-Berlin.

Some doubts about Nachtsheim's sincerity are appropriate when he equates the conditions of research in Nazi Germany with those in the USSR. An analysis of his research during National Socialism does not reveal any restrictions in scientific freedom. This holds true not only for Nachtsheim. Biological research on the whole, and genetic research in particular, flourished until 1944 without noticeable restrictions for those who were neither Jewish nor outspoken anti-Nazis.[24] I am not speaking here about research in race hygiene carried out mainly by medical scientists and anthropologists but of basic genetic research by biologists. Nachtsheim was well funded. His grant application of 1935 at the DFG/RFR was refused, but he was funded continuously from 1937 on. Funding increased strongly after his move to the KWI for Anthropology. Von Verschuer received altogether 80,000 RM in 1943 and 1944 for experimental hereditary pathology, most of which was for Nachtsheim's research. From 1933 through 1945, Nachtsheim published 48 papers and one book.

In 1933, Nachtsheim decided to change his research in order to serve the race hygienist aims of the Nazis, aims which he publicly supported; nobody forced him to do so. Other geneticists continued to conduct research in *Drosophila* or *Antirrhinum*. Although some Nazi politicians were hostile toward the pure sciences and demanded that science serve the ideological and practical ends of the regime,[25] others were interested in having outstanding scientists in leading positions, particularly those from the KWI, and in having their research funded.[26] Freedom of science, though not propagated officially, seems to have been the case at least in major parts of biology during Nazi Germany.

Here I detect a difference between the science policy and terror of Nazi Germany and that of Stalinist USSR: Stalin used terror and mass murder to achieve political aims. His terror could affect nearly everybody; apart from the Kulaks, it was not confined to a particular group. For science that meant that no one could be certain he was exempt from the terror, even if he pretended to adopt Lysenko's views. Moreover, the strong politicization of biology that resulted from Lysenko culminated in the notorious conference of the Lenin Academy in 1948, which curtailed freedom of research almost completely.[27]

In Nazi Germany, on the other hand, most biologists could work in relative security after the wave of terror and expulsions of the early 1930s. The vast majority of those dismissed between 1933 and 1935 were Jewish or of Jewish origin, and they were dismissed regardless of their political opinions or the content of their research. Of all the biologists who stayed in Nazi Germany after 1933, I know of only one case,

Walther Arndt, who was murdered by the regime. He was denounced in 1943 for publicly doubting the final German victory.

Having said this, I want to emphasize that by no means do I want to diminish the Nazi terror or deny the Holocaust, the center of the terror, the systematic annihilation of the European Jewry.[28] I am just saying that the impact of Stalinism and National Socialism on biology seem to have been different. And in searching for the reason for the relative security of ordinary scientists in Nazi Germany, one must ask, why wasn't there any criticism of the regime? Why didn't any geneticists, even those who were not like Fischer or von Verschuer racists themselves, criticize the anti-Semitic, racist measures of the state that were carried out in the name of genetics? One answer – it could have been the end of their career, and they profited from this regime – has already been acknowledged. Scientists paid for freedom of research in their laboratories by remaining silent about its application or its misuse.

In many instances, freedom of research meant more than the right to choose one's research topic. It is well known that von Verschuer started a close collaboration with his former assistant Mengele in Auschwitz in 1943. Mengele sent him "material," i.e., blood, eyes, and other organs from murdered people.[29] In a committee of professors who met later to decide on the possible guilt of von Verschuer, Nachtsheim criticized him for his open racism but objected about von Verschuer's knowing about the unnatural deaths of the corpses he used in his experiments. In a dispute with the physicist Robert Havemann, who had attacked von Verschuer for having collaborated with Mengele, Nachtsheim defended the use of human "material" from concentration camps if a person was to be killed anyway.[30] Nachtsheim believed that using the corpses of people executed for political reasons (as in the case of Prof. Hermann Stieve) for "valuable research which would not have been possible otherwise" was similar to using human "material" from concentration camps.[31] In his view, von Verschuer should be reproached only if he had known that the people whose eyes he analyzed had not died from a natural death or that they had been killed for that purpose. Von Verschuer denied such knowledge, how could one prove it? I should add that the eyes came from members of the same family, several of them twins.

Hermann Stieve, an anatomist, tried to demonstrate the influence of psychic factors on ovulation. Thus he contradicted the gynecologist Hermann Knaus, who claimed that ovulation would always occur at a certain day of a woman's cycle. Stieve confirmed his theory of paracyclic ovulations by using the corpses of healthy young women who had been sentenced to death.[33] Such studies show, among other things, that human geneticists and other medical scientists during the Nazi era enjoyed a freedom that they neither had before nor afterwards.

In exculpating von Verschuer, Nachtsheim contributed substantially to the absolution of human genetics from its worst crimes. And he went even further. In March 1945, shortly before the Allies arrived in the city, von Verschuer had already left Berlin in order to spend the last weeks of the war in Solz, a more secure place. Nachtsheim wrote a very friendly letter to him there. He was worried about the documents that von Verschuer had left in Berlin; according to him, they should have been destroyed but were not. Nachtsheim offered his help: "...I learned from Miss Jarofki that a large amount of documents have remained here, which in the case of an occupation by the enemy (*"im Falle eines Feindeinbruchs"*) should be destroyed. I have not yet found out which material and how much is relevant, but I assume that Miss Jorafki is well informed. You didn't discuss this with me; otherwise, I would have advised you to take the stuff with you to Solz. In any case, we should not wait too long to destroy the material, and I believe I could take the responsibility for making the appropriate decisions...."[33]

In a letter to L.C. Dunn of February 14, 1961, Nachtsheim made it clear that he had known already in the early 1940s about the unnatural death of these people from Auschwitz: "I must confess that it was the biggest shock that I experienced during the whole Nazi period when one day Mengele sent me the eyes of a Gypsy family that was housed in the concentration camp Auschwitz."[34] To conceal being himself a witness of murdering for science might be one reason why Nachtsheim helped destroy the incriminating material and exculpated von Verschuer. Another reason might be the connection to his own research. Stieve wanted to find out about the paracyclic ovulation, and he took advantage of "human material" that fell victim to the NS regime. There is evidence that at least in some cases Stieve could determine the day of the murder so that the "material" would fit his research.[35] Von Verschuer wanted to know about the racial differences of blood proteins, and he benefited from his collaboration with Mengele, who sent the required "material." Nachtsheim wanted to find out how to diagnose hereditary epilepsy. He, too, made use of people deprived of their rights for his own experiments. So the question arises, in exculpating others, did Nachtsheim also want to exculpate himself? Contrary to what Nachtsheim claimed after 1945, in human genetics it was not freedom of science that was lacking during National Socialism but, rather, respect for the dignity of the individual.

Acknowledgements

I thank Diane Paul, Otto Geudtner, Benno Müller-Hill, and Niels Roll-Hansen for stimulating discussions and critical comments, and Eddie Goldberg for his critical reading of an earlier draft of the manuscript. The work was supported by grant Mu 575/4 of the Deutsche Forschungsgemeinschaft through *Schwerpunkt Wissenschaftsemigration*.

Notes and References

1. Hermann Muller to Max Delbrück, February 24, 1947, Delbrück papers, California Institute of Technology.
2. Paul Weindling, *Health, Race, and German Politics Between National Unification and Nazism, 1870 –1945*, (Cambridge, England: Cambridge University Press, 1989), p. 566.
3. Hans Nachtsheim to Hans Grüneberg, April 2, 1948, Grüneberg papers, ICHM London; personal communication by Jonathan Harwood.
4. Peter Weingart, Jürgen Kroll, and Kurt Bayertz, *Rasse, Blut und Gene – Geschichte der Eugenik und Rassenhygiene in Deutschland* (Frankfurt: Suhrkamp, 1988), p. 583.
5. The *Habilitation* is an academic degree beyond the doctorate, which allowed the holder to teach at a university.
6. Personal communication by Jonathan Harwood, August 1993.
7. *Ibid.*
8. *Gutachten der Dozentenschaft betr.* n.b.a.o. Prof. Dr. Nachtsheim, November 17, 1936, Archiv Humboldt-Univ., ZGI/742. (After the *Gleichschaltung* of the universities in 1933, the Reich education minister based his decisions about positions on expert opinions from the faculty and the Nationalsozialistische Deutsche Arbeiter Partei; they dealt with the scientific expertise, as well as the political attitude of the person.)
9. Hans Nachtsheim to Deutsche Forschungsgemeinschaft, May 14, 1937, Bundesarchiv Koblenz, R73/13328. Nachtshein wrote that he had successfully started *"ein neues, auch für die menschliche Rassenhygiene wichtiges Gebiet."*
10. Diane B. Paul, "The Cold War in Genetics: Hans Nachtsheim and Human Genetics in Post-War Germany," address given at the History of Science Annual Meeting, Santa Fe, New Mexico, November 12, 1993. I didn't get access to the Nachtsheim papers in the Archives of the Max-Planck-Society and therefore thank Diane Paul for this information.
11. Hans Nachtsheim, 'Erbleiden beim Tier in ihrer Bedeutung für die menschliche Erbpathologie," *Naturwissenschaften, 32* (1944), pp. 348 – 361.
12. Gerhard Ruhenstroth-Bauer and Hans Nachtsheim, "Die Bedeutung des Sauerstoffmangels für die Auslösung des epileptischen Anfalls," *Klinische Wochenschrift, 23* (1944), pp. 18 – 21.
13. Hans Nachtsheim to Deutsche Forschungsgemeinschaft, March 15, 1944, Bundesarchiv Koblenz, R 73/15342.
14. Gerhard Ruhenstroth-Bauer and Hans Nachtsheim, "Die Bedeutung des Sauerstoffmangels für die Auslösung des epileptischen Anfalls," *Klinische Wochenschrift, 23*

(1944), pp. 18 – 21.

15. Benno Müller-Hill, "Genetics after Auschwitz," *Holocaust and Genocide Studies, 2* (1987), pp. 3 – 20.

16. Gerhard Koch, *Humangenetik und Neuro-Psychiatrie in meiner Zeit (1932 –1978) – Jahre der Entscheidung* (Erlangen: Palm & Enke, 1993), p. 125.

17. This and all subsequent translations from the German are my own.

18. Ernst Klee, *"Euthanasie" im NS-Staat – Die "Vernichtung lebensunwerten Lebens"* (Frankfurt: Fischer, 1983), p. 300.

19. H.-H. Knaape to B. Müller-Hill, 5 December, 1988

20. Hans Nachtsheim, *Für und wider die Sterilisation aus eugenischer Indikation*, Stuttgart 1952, p. 50.

21. Alexander Mitscherlich, "Eugenik – Notwendigkeit und Gefahr," *Fortschr. Med., 81* (1963), pp. 714 –715.

22. Hans Nachtsheim, "Bourgeoise Biologie," *Der Tagesspiegel*, August 5, 1948.

23. Hans Nachtsheim, "Lyssenkos Ende," *Deutsche Kommentare*, April 14, 1956.

24. See my larger study, *Biologists*, chapter 2 – 4.

25. See, for example, Joseph Needham, *The Nazi Attack on International Science*, (London: Watts & Co., 1941).

26. Ute Deichmann and Benno Müller-Hill, "Biological Research at Universities and Kaiser Wilhelm-Institutes in Nazi Germany," *Science, Technology and National Socialism*, ed. Monika Renneberg and Mark Walker (Cambridge, England: Cambridge University Press, 1994), pp. 160 –183.

27. The impact of Lysenko on the biosciences in the USSR has been analyzed, for example, by David Joravsky, *The Lysenko Affair*, reprint (Chicago: The University of Chicago Press, 1970).

28. For comparison of Hitler's and Stalin's terrors, see, for example, Alan Bullock, *Hitler and Stalin. Parallel lives* (London: Harper Collins, 1991).

29. Benno Müller-Hill, *Murderous Science. Elimination by Selection of Jews, Gypsies and Others, Germany 1933 –1945* (Oxford: Oxford University Press, 1988), pp. 70 –74.

30. Weindling, *Health, Race*, p. 569; Weingart, Kroll and Bayertz, *Rasse, Blut*, p. 575. I didn't get access to the Nachtsheim papers in the Archives of the Max-Planck-Society, which contain the correspondence with Havemann.

31. See Weingart, Kroll, and Bayertz, *Rasse, Blut*, pp. 572 – 581.

32. Hans Harald Bräutigam, "Tod nach Kalender," *Die Zeit*, January 20, 1989.

33. Hans Nachtsheim to Otmar von Verschuer, March 12, 1945; I thank Raphael Falk for having shown this letter to me, cited in his unpublished manuscript "Hans Nachtsheim: How to be a Eugenicist in National Socialist Germany and Prevail."

34. Hans Nachtsheim to L.C. Dunn, February 14, 1961 (author's translation), in Paul, "The Cold War in Genetics...."

35. Hans Harald Bräutigam, "Tod nach Kalender," *Die Zeit*, January 20, 1989.

GOOD GENES AND BAD GENES

DNA in Popular Culture

DOROTHY NELKIN

Dept. of Sociology, New York University, 269 Mercer Street, New York, NY 10003, USA

M. SUSAN LINDEE

Department of History and Sociology of Science, University of Pennsylvania, Philadelphia, PA, USA

In popular biographies, Elvis Presley appears as a genetic construct, driven by his genes to his unlikely destiny. He has succeeded, the story goes, because of his genetic heritage – and failed because of his family's history of inbreeding. Elaine Dundy, for example, attributes Presley's success to the qualities of will, ambition, and fantasy passed down to him from his mother's multi-ethnic family.[1] Dundy traces Elvis's musical talents to his father who "had a very good voice" and his mother who had "the instincts of a performer." They did provide a musical environment, she notes, but "even without it, one wonders if Elvis, with his biological musical equipment would not still have become a virtuoso."

Another Elvis biographer, Albert Goldman, focuses on his "bad" genes, describing him as "the victim of a fatal hereditary disposition."[2] Recalling early twentieth-century descriptions of families with degenerate traits, Goldman attributes Elvis's character to his ancestors, who constituted "a distinctive breed of southern yeomanry," commonly known as hillbillies. A geneology research organization, Goldman reports, has traced his family tree back nine generations to a nineteenth-century ancestor who was a "coward, deserter, and bigamist." In Goldman's narrative, this heritage was the key to Elvis's downfall: his addiction to drugs and alcohol, his emotional disorders, and his premature death could all be explained by his genetics. His fate was a readout of his DNA.

This paper includes material from our book, *The DNA Mystique; The Gene as a Cultural Icon* (New York: W.H. Freeman, 1995). Research was supported by the NIH, Grant 1RO1 HG 0047-01.

Michael Fortun and Everett Mendelsohn (eds.), The Practices of Human Genetics, 155–167
©1999 *Kluwer Academic Publishers. Printed in Great Britain.*

Many popular stories these days explain "good" and "bad" behaviors as written in the genes. The body in these stories is a transcription of an underlying text; attributes are self-evident signs of an inner code; human fate reflects hidden genetic predispositions. The stories of bad genes commonly focus on criminal behavior, war, or alcoholism. Stories of good genes are about success, special talents, or unusual skills. These narratives are fatalistic, suggesting that individuals succeed or fail not because of their efforts, their wills, or their social circumstances, but because they have been programmed genetically for their fate. This paper, developed from our study of representations of the gene in popular culture, explores this form of fatalism.

Genetics has entered the vernacular over the past decade as a prominent cultural image. Magazine articles, television stories, advertisements, jokes, films, novels, and childcare books have appropriated the gene as a metaphor or plot device, defining the person as a DNA readout. In part, popular interest in the gene reflects the visibility of the science of genetics. But our study suggests that genetic images also appeal because they seem to offer compelling and simple explanations for complex and enduring human questions. Thus our goal is to understand the gene as a cultural icon, a way to make sense of the world. And our interest is in the implicit pedagogy that emerges from repeated images and narratives, as well as its effect on social policies and popular expectations.

Evil in the Genes

Contemporary narratives of popular culture often attribute behavior that threatens the social contract to "bad genes" – to the innate, genetic constitution of individuals. Evil, according to one journalist, is "embedded in the coils of chromosomes that our parents pass to us at conception."[3] Based on this assumption, Oprah Winfrey, in a live interview on her prime-time show, found it meaningful to ask a twin whether her sister's "being bad" is "in her blood?"[4] In the movie *JFK* John Candy told Kevin Costner, "You're as crazy as your mama – Goes to show it's in the genes."[5] And bad genes have become a facetious metaphor to describe national aggression. A *Time* article on the "New Germany" described the nation as "a child of doubtful lineage adopted as an infant into a loving family; the child has been good, obedient, and industrious, but friends and neighbors are worried that evil genes may still lurk beneath a well-mannered surface."[6]

The existence of evil has posed problems for philosophers and theologians for much of human history. Evil can be seen as the cosmic consequence of fate, or as the result of voluntary human action or moral failure. The agents invoked to explain the presence of evil are commonly powerful and somewhat abstract – demons, gods, witches, a marked soul, and now, the biochemistry of the brain. Genetic explanations of "evil" are one form of a long-standing belief that individuals do not entirely control their own actions. The belief that the "devil made me do it" does not differ significantly in its consequences from the belief that "genes made me do it." Both explanations locate control over human fate in powerful, abstract entities capable of dictating human action in ways that mitigate moral responsibility and alleviate personal blame.

Genes are frequently used to explain a common and troubling contradiction. Why do some individuals, despite extremely difficult childhoods, become productive, even celebrated members of society, while other children who were granted every opportunity and advantage turn out badly? In December 1991, a 14-year-old high school boy was arrested for the murder of a schoolmate. The *New York Times* account of this event interpreted it as a key piece of evidence in "the debate over whether children misbehave because they had bad childhoods or because they are just bad seeds...." The boy's parents had provided a good home environment, the reporter pointed out. They had "taken the children to church almost every Sunday, and sacrificed to send them to a Catholic grammar school." Yet their son had been arrested for murder. This troubling inconsistency between the child's decent background and his violent behavior called for explanation. The reporter resolved the mystery through the explanatory power of inheritance: the moral of the story was stated clearly in its headline: "Raising Children Right Isn't Always Enough."[7]

A 1993 prime-time television play called *Tainted Blood* drew on the same set of ironies.[8] A 17-year-old boy from a good family killed his parents and then himself, shocking the community. The case attracted an investigative reporter (Raquel Welch) who found out that the boy had been adopted while his genetic mother was in a mental institution. Suspecting he might have "inherited the gene for violence," the reporter went to the institution, where an elderly doctor (portrayed as out-of-date) insisted that her ideas about heredity were wrong and that teenagers who murder were invariably abused. But in the film, it was the doctor who was wrong. The reporter discovered that the mother, like the son, had killed her parents and then herself. Moreover, the boy had a twin sister who had also been adopted. Thus there began an urgent search for the girl with "tainted blood." She, too, was predisposed to violence and, after threatening her family, killed herself. But, said the reporter, she was not to be blamed for her actions, because she had inherited a "genetic disease:" she was "born to kill."

These and many other stories set up a conflict between childhood experience and adult behavior. And when the two seem to conflict, hereditary predisposition – the force of genetics – provides a plausible and appealing resolution.

Research that indicates a family link to criminality attracts extensive media attention. The news coverage of research on a Dutch family found to share a genetic mutation that seemed to predispose its males to aggressive behavior sounded much like the reports of pathological families in the early part of the century. Though journalists will often qualify, indicating that family patterns do not settle the debate over root causes, they continue to promote the theories of those scientists who believe that genetic predisposition is an underlying basis of violent behavior.[9] A PBS series on "The Mind" introduced a segment on violence: "Recent research suggests that even the acts of a serial killer may have a biological or genetic basis."[10] Donahue described research on the extra Y chromosome as a way for parents "to tell if your child is a serial killer."[11] He and other media personalities refer to "criminal genes," as if such genes can be isolated, identified, and used to predict particular actions.

The media interest in biological explanations of violence reflects the hope that genetic information will enable the prediction, and, therefore, the control of deviant behavior. Scientists encourage such expectations. In a *Science* editorial, the journal's editor, biologist Daniel Koshland, tells stories about recent acts of violence: "An elephant goes berserk at the circus, an elderly pillar of the community is discovered to be a child molester, a man admits to killing many young boys..., a disgruntled employee shoots seven co-workers." Each crime, Koshland says, has a common origin – an abnormality of the brain.[12] Psychologist Jerome Kagan, often quoted in the media for his work on the hereditary basis of shyness, also talks to reporters about violence. He expects that within 25 years genetic tests will be able to pick up about 15 out of every 1,000 children who may have violent tendencies, though only one of these will actually become violent.[13]

Even when scientists emphasize that violence reflects a mix of biological and environmental causes, media accounts highlight the importance of genetics. This tendency was demonstrated in the press coverage of the 1992, National Research Council review of the state of research on violence.[14] The report said that violence arises from the "interactions among individuals' psychosocial development, neurological and hormonal differences, and social processes." It described the studies used to demonstrate genetic influence on individual potential for anti-social behavior as having highly uncertain implications. While not ruling out genetic processes, the report suggested: "If genetic predispositions to violence are discovered, they are likely to involve many genes and substantial environmental interaction rather than any simple genetic marker." Only 14 of the 464 pages of the NAS report actually dealt with biological perspectives on violence, and less than 2 pages were about genetics. Nevertheless, the *New York Times* headlined its article, "Study Cites Role of Biological and Genetic Factors in Violence."[15] Genes are far more newsworthy than social or economic circumstances as a source of antisocial behavior. The source of their appeal is suggested by the language of a recent article in *Time*, looking for the causes of "the savagery that is sweeping America." It suggested that society's ills can-not fully be responsible, that violence may be caused by "errant genes." "Science could help shed light on the roots of violence and offer new solutions for society."[16]

Biological theories also appeal as explanations for war. In the early 1970s, ethologists Marvin Harris, D.P. Barash, and Konrad Lorenz began to promote a biological model of organized human aggression, explaining war as a productive and necessary social activity.[17] Their books, published for a lay audience, became fashionable, attracting extensive media cover-age. Reviewing their perspective on aggression research, Jeffrey Goldstein, a psychologist from Temple University, found that the media have systematically covered studies that offer evidence of genetic explanations of violence, and are less interested in research on the influence of social and economic conditions.[18]

Biologists and social scientists have criticized research on the genetic predisposition to organized aggression for concealing inadequate method-ologies behind quantitative data, and for minimizing the well-studied effect of social environment. In May 1986, Goldstein helped assemble a group of these critics to discuss biological theories about the origin of warfare. Meeting in Spain, they produced the Seville Statement on Violence, which strongly repudiated the idea that war is biologically necessary or genetically controlled. "It is scientifically incorrect to say that war is caused by 'instinct' or any single motivation...scientifically incorrect to say that humans have a 'violent brain'...scientifically incorrect to say that in the course of human evolution there has been a selection for aggressive behavior more than for other kinds of behavior...." The statement concluded that "biology does not condemn humanity to war...The same species who invented war is capable of inventing peace."[19]

This brief but unambiguous text was signed by 20 well-known scholars from around the world and endorsed by the major professional associations. Yet, despite considerable efforts to publicize the statement, it attracted little media attention. Reporters, writes Goldstein, seem to prefer explanations of aggression that invoke biological necessity. A journalist, responding to the efforts to disseminate the Seville material expressed the prevailing attitude, "Call me when you find the gene for war."[20]

The interest in "bad genes" is also evident in scientific and social speculation about the nature and etiology of addiction. Definitions of alcoholism have shifted from sin to sickness, from moral transgression to medical disease, depending on prevailing social agendas.[21] Common observation shows that alcoholism runs in families. As in the case of violence, however, this in itself tells us little; for family continuities could reflect role models, the availability of alcohol, or the reaction to abuse. Nevertheless, the prevailing perception is that expressed in *Omni*: "Addicted to the bottle? It may be in your genes."[22] The gene for alcoholism became a talk show theme on the Oprah Winfrey and Phil Donahue shows. An article in *Mademoiselle* asks: "Do you have a gene that makes you a designated drinker?" and suggests that "even if you have exceptional self-discipline, you could still be at high risk."[23] And Nancy Reagan's, famous antidrug slogan "Just say no" evoked comments about the "genes-impelled compulsion" to take drugs.[24]

In an article on addiction, *New York Times* reporter Dan Goleman presented several cases to dramatize the genetic basis of alcoholism.[25] A 26-year-old executive was the class clown as a child and class president in his high school. Always extroverted, as he matured he started taking drugs to stay high. Addiction appealed because of his "natural bent." Another young man had been anxious as a child until he discovered that alcohol made him relax. His father was an alcoholic so he had access to liquor, but Goleman explained his addiction in terms of biological vulnerability, citing a scientist who claimed that genetic engineering will eventually eliminate the gene.

This is one reason for the appeal of genetic explanations. In effect, like genetic explanations of violence, identifying the "alcohol gene" – or, in other stories, the gene for smoking, overeating, shopping, or gambling – offers a window of hope for controlling addiction, not through the uncertain route of social reform but through biological manipulation. However, causal explanations also imply moral judgments about responsibility and blame. If defined as a sin, alcoholism represents an individual's flaunting of social norms; if defined as a social problem, it represents a failure of the social environment; if defined as intrinsic to the product, it represents the need for regulation. But if defined as genetic, neither the society nor the industry are responsible, and even the addicted individual cannot really be blamed.

The definition of addiction and, more generally, of deviant behavior remains a loaded question. To explain human behavior in absolute genetic or biological terms is to extract it from the social setting that defines and interprets behavior. There are clearly no criminal genes, or alcohol genes, but only genes for the proteins that influence hormonal and physiological processes. As biologist Richard C. Lewontin points out, only the most general outline of behavior can be genetically coded.[26] Thus, while genes may indirectly affect behavior in particular social circumstances, they are hardly deterministic. Indeed, defining addiction, crime, and war as molecular events writ large reflects cultural not scientific interpretations. This cultural meaning of the gene is evident when "good genes" appear in descriptions of positive traits to explain special talents, the success of celebrities, and even the qualities of inanimate objects like automobiles, perfumes, or magazines.

Good Genes

Sixteen-year-old Judit Polgar is the youngest chess grand master ever and the first female chess player considered to have the potential to be a world champion. Her parents insist that her talent is a consequence of family training. She and her two sisters, all world-class chess players, have never attended school, but are tutored at home by their multilingual mother and their psychologist father, who believes that "every child is a potential genius." He raised his children as an "experiment" in the power of environment, and by all counts it is a successful experiment: Judit and her sister Zsuzsa are grand masters, ranked respectively first and second female chess players in the world, and their middle sister, Sofia, is ranked sixth. But according to a *New York Times* account of the remarkable Polgar sisters, their father's environmental explanation is "looked upon with skepticism" in the chess world. Another grand master is quoted as saying, "I think they were born gifted, and one is a genius."[27]

There are many possible explanations for success; hard work, persistence, the exposure to role models, contacts and professional opportunities, social pressures from family or peers, or simply random good luck. Clichés suggest the great variety of traditional explanations: "Practice makes perfect," "Seek and you will find," "He pulled himself up by his bootstraps," "He had it made," or "The rich get richer." The Horatio Alger myth underlying American narratives of success suggests that any person who tries can "make it." Ubiquitous stories describe the "self-made man." The myth of the "Jewish mother" suggests the role of guilt in the Jewish family as a driving force in the children's success. And the recent success of Asian students has evoked stories about the importance of parental pressure.

But another set of clichés – "Good blood," or "Blood will tell" – suggests that success is biologically determined. Just as children from good families may turn out bad, so those with limited opportunity can rise above their circumstances because of their genes. Thus a television newscaster describes a teenager named Mike who, though raised in a poor family with no father, becomes captain of his track team and wins a college scholarship: "He has a quality of strength, and I guess it has a genetic basis."[28] A *Newsweek* article asserts that "Some kids have protective factors that serve as buffers against the risks." They have "natural resilience" or "built-in defenses." It is the "genetic luck of the draw."[29]

Genetic explanations of socially valued traits frequently appear in stories about popular personalities. When talented people – scientists, actors, sports heroes, politicians, or rock musicians – become media stars, their genes, as well as their genius, are exposed to speculation. We read about Elvis's genes and Einstein's DNA. In a culture obsessed with fame, money, and personal success, the location of exceptional achievement is a matter of widespread curiosity. Why are some more successful than others? What accounts for extraordinary achievement? The tendency to locate success in the DNA is part of the growing cultural power of genetic explanations. It implies a significant change in the "bootstrap" ideology that has pervaded American folklore; for genetic explanations undermine the importance of hard work as the route to success. Neither individual actions nor social opportunity matter if our "fate" lies in our genes.

Popular profiles of famous people describe their genes without benefit of molecular biology. The special skills, the striking talents, and the complex behavior of public figures appear to be passed down through the generations like brown hair or blue eyes. The range of special talents attributed to genetics is remarkable. In a story about the Ginsberg brothers, both of whom are poets, the *New York Times* refers to their "poetry genes."[30] An obituary writer explains the secret of Isaac Asimov's success: "Its all in the Genes."[31] A business writer describes entrepreneurs as having inherited business tendencies.[32] A fashion columnist in the *New York Times* calls her story: "Fashion's Nature vs. Nurture Debate, Or Is Good Taste in the Genes?"[33] There have been many definitive and detailed biographies of James Joyce, but one more appeared in 1993. To the wealth of details known about Joyce's life, Peter Costello's biography includes extensive family trees and a diagram of Joyce's "genetic make-up."[34] Even Mother's Day cards use quips about genetics, for example, imperfections "must be the genes from Dad's side of the family."

Inanimate objects can also be successful by virtue of their genes. For example, the gene is a popular image in automobile advertising. A Sterling's remarkable handling is "in its genes." A BMW sedan has "a genetic advantage." A Toyota, says a pregnant woman in an ad, has "a great set of genes." Apparently, other products have good genes as well. A Bijan perfume is called "DNA," and the company advertises it as "a family value" and "the stuff of life."[35] A blue jeans ad exploits the obvious pun "Thanks for the genes, Dad" and implies their superior quality.[36] An article on the leadership changes in the *New Yorker* asks, "Can you change a magazine's DNA?...A magazine's underlying character remains – unchanged and enduring, a DNA-like set of fingerprints – and lasts through the years and reinventions....Tina Brown has much to reckon with, starting with 67 years of DNA."[37]

Thus "good genes" as a metaphor and an explanation appear in the popular culture to describe consumer products, as well as people. And sometimes people themselves become products. A Maybeline commercial features the well-known fashion model Christy Turlington. Two questions are flashed on the screen: "Is it in her Genes?" and "Was she born with it?"

The media are attracted to celebrity families with similar professional interests. Children often do enter the professions of their parents, and they are attracted to these professions for many reasons. Social expectations, parental pressures, the availability of unique opportunities, and personal contacts may all influence the selection of careers. But many accounts of children who follow the career paths of their parents talk about their genes. As in the earlier narratives of the eugenics movement, a trait that is shared by both parent and child is frequently assumed to be genetic. When Ringo Starr, former Beatles drummer, was interviewed on the Arsenio Hall Show, Arsenio commented on his son. "He's a drummer too? He must have the drummer gene."[38] A host of the "Today Show" introduced the daughter of the singer Marvin Gaye as having her father's "talent genes."[39] And in the CBS coverage of the 1992 Olympic ice skating competition, the announcer explained the success of the Japanese skater Yuko Sato, "She has a genetic advantage" because her parents were skaters too.[40] Celebrities sometimes encourage genetic explanations. Michael Jackson, describing his own talents, compares the "biological rhythms that sound out the architecture of my DNA" to the "life songs of ages" and the "music that governs the rhythms of the seasons...."[41]

Even William Safire, though sharply critical of the casual misuse of language, has used genetic metaphors. In a *New York Times* editorial about the romance between Albert Einstein and Miliva Maric, he speculated about their daughter who had been given up for adoption, "We can presume she grew up to have a family of her own and that humanity has been enriched by the propagation of the genes of genius."[42]

Safire's comment reflects the considerable professional and popular curiosity about the origins of scientific genius. In an article called "Love of Science: Do Parents Pass it Along to Their Children?" a reporter interviewed experts on the effect of genetics on people's interest in science. One expert conveyed the prevailing wisdom, "It is more socially correct to think of everyone as a blank slate...[but] my inclination is to think, yes, genetics are involved."[43]

Genetics has become a politically correct way to characterize the foibles of successful politicians. Thus a journalist described former President George Bush as missing an empathy gene.[44] Another suggested that presidential candidate Ross Perot inherited "his father's frugal gene and no amount of lucre would change that." (Perot had insisted that his children take their own popcorn to the movies rather than waste their allowance on overpriced theater concessions.)[45] When Pat Buchanan was running as a presidential candidate, a reporter referred to one of his aides as a "genetic conservative."[46] Before the 1992 presidential election, a political joke suggested that Democratic men in Washington were dating Republican women to replenish their gene pool, so they could produce a winner.[47]

The appropriation of DNA – the good or bad gene – to explain individual differences conveys a moral message. The great, the famous, the successful, are what they are because of their genes. This becomes a way to neutralize success, to level celebrities. They cannot claim moral superiority; they are no better than the rest of us, for their fate lies in their genes. But there is also a social message. Opportunity is less important than predisposition. The star – or the criminal – is not made but born. Some people are destined for success, others for problems or, at least, a lesser fate. This is a particularly striking theme in American society, where the very foundation of the democratic experiment was premised on the belief in the improvability, indeed, the perfectibility of all human beings. But it is a popular theme in the 1990s. Witness the extraordinary media receptivity to *The Bell Curve*, a book that relates social and economic status in America to genetically based differences in IQ and makes claims for its relevance to social policies.[48]

The belief in genetic destiny implies that there are natural limits constraining the possibilities for both individuals and social groups. Humankind is not perfectible because the species's flaws and failings are inscribed in an unchangeable text – the DNA – that will persist in creating murderers, addicts, the insane, and the incompetent, even under the most ideal social circumstances. In popular stories, children raised in ideal homes become murderers, and children raised in difficult home situations become well-adjusted high achievers. The moral? There is no possible ideal social system, no possible ideal nurturing plan that can prevent the violent acts that seem to threaten the social fabric of contemporary American life.

At the same time, some children are destined to greatness – to fame, fortune, or political leadership – regardless of how they are raised or what obstacles they face. Social reforms can have only limited consequences, since those with the right characteristics will succeed, no matter what. From both perspectives, the idea of genetic predisposition (for success or for failure) bears on the location of responsibility and blame.

As our analysis suggests, the gene in popular culture is more than a biological entity. It is a cultural resource that can be invoked to explain nearly every personality and behavioral trait. It is a political resource that can effectively absolve society – and even the individual – of responsibility for behavior. Genetic determinism appeals in many policy contexts, therefore, as a justification for passive attitudes toward social injustice, and even aggressive neglect of continuing social problems. And it implies a dangerous way out of social dilemmas – suggesting that the improvement of society depends, ultimately, on the improvement of DNA.

Notes and References

1. Elaine Dundy, *Elvis and Gladys* (New York: St. Martin's, 1985). p. 26. See also discussion in Greil Marcus, *Dead Elvis: A Chronicle of a Cultural Obsession* (New York: Doubleday, 1991).
2. Albert Goldman, *Elvis*, (New York: McGraw-Hill, 1981). p. 57. See also discussion in Greil Marcus, *Dead Elvis*, 1991.
3. Deborah Franklin, "What A Child is Given," *New York Times Magazine*, September 3, 1989. p. 36.
4. "Oprah Winfrey Show," *CBS*, August 24, 1992.
5. *JFK*, Warner Bros, 1991.
6. James O. Jackson, "The New Germany Flexes Its Muscles," *Time*, April 13, 1992. p. 34.
7. Maria Newman, "Raising Children Right Isn't Always Enough," *New York Times*, December 22, 1991.
8. *Tainted Blood*, USA Channel, March 3, 1993.
9. Fox Butterfield, "Studies Find a Family Link to Criminality," *New York Times*, January 31, 1992.
10. Richard Hutton and George Page "The Mind/The Brain Classroom Series," *PBS Video*, 1992.
11. Donahue Show, February 25, 1993. The program was described in John Horgan, "Eugenics Revisited," *Scientific American*. June 1993, p. 123.
12. Daniel Koshland, "Elephants, Monstrosities, and the Law," *Science, 255*, February 14, 1992, p. 777.
13. Quoted in Anastasia Toufexis, "Seeking the Roots of Violence," *Time*, April 19, 1993, pp. 52–3.
14. National Academy of Sciences, National Research Council, *Understanding and Preventing Violence*, National Academy Press, November 1992.
15. Fox Butterfield, *New York Times*, November 13, 1992.
16. Toufexis, "Seeking the Roots...," 1993.
17. See Marvin Harris, *Cows, Pigs, Wars and Witches: The Riddle of Culture* (New York: Random House, 1974); D.P. Barash, *The Whisperings Within* (New York, Harper and Row, 1979); and Konrad Lorenz, *On Aggression* (New York: Bantam, 1967).
18. Jeffrey Goldstein, *The Seville Statement on Violence*, November 1990.
19. The Seville Statement and list of signatories is included in Anne E. Hunter, ed., *Genes and Gender VI: On Peace, War and Gender* (New York: The Feminist Press, 1991) pp. 168–171.
20. Jeffrey Goldstein, *The Seville Statement on Violence*, November 1990, p. 41.
21. Sheila B. Blume, M.D., "The Disease Concept of Alcoholism, 1983," *Journal of Psychiatric Treatment and Evaluation, 5*, pp. 417–478. She traces the modern conception of alcoholism as a disease back to Benjamin Rush and notes its subsequent history.
22. George Nobbe, "Alcoholic Genes," *Omni*, May 1989. p. 37.
23. Shifra Diamond, "Drinking Habits May be in The Family," *Mademoiselle*, August 1990. p. 136.
24. Editorial, "Just Blame Genes – of Disease," *Christian Science Monitor*, May 22, 1991.

25. Daniel Goleman, "Scientists Pinpoint Brain Irregularities In Drug Addicts," *New York Times*, June 26, 1990.
26. Richard C. Lewontin, *Biology as Ideology* (New York: Harpers, 1992) p. 51.
27. Bruce Weber, "Chess Moves are Planned, Birthdays Happen," *New York Times*, August 5, 1992.
28. NBC News Special, "Kids and Stress," April 25, 1988.
29. David Gelman, "The Miracle of Resiliency," *Newsweek* Special Issue. Summer 1991. pp. 44–47.
30. Barbara Delatiner, "For Brothers, Poetry is in Their Genes," *New York Times*, May 26, 1991.
31. Mervyn Rothstein, "Isaac Asimov, Whose Thoughts and Books Traveled the Universe, Is Dead at 72," *New York Times*, April 7, 1992, p. B7.
32. Diane Cole, "The Entrepreneurial Self," *Psychology Today*, June, 1989, p. 60.
33. Maria Terrone and Sharon Johnson, "Fashion's Nature Vs. Nurture Debate Or, Is Good Taste in the Genes?" *New York Times*, April 12, 1992, advertising section.
34. Peter Costello, *James Joyce: The Years of Growth* (New York: Pantheon, 1993); Christopher Lehmann-Haupt observed the focus on genetics in a review in the *New York Times*, April 8, 1993.
35. *Mirabella*, January 1993. The bottle pictured in some ads for Bijan's DNA has the amazing shape of a triple helix.
36. Calvin Klein ad, quoted in Anne Fausto-Sterling, *Myths of Gender* (New York: Basic Books, 1985). p. 7.
37. Edwin Diamond, "Can You Change a Magazine's DNA?" *New York Magazine*, July 20, 1992. p. 27.
38. "Arsenio Hall Show," Fox Television Network, August 2, 1992.
39. "Today Show," NBC, October 21, 1992.
40. Scott Hamilton, "Olympic Women's Ice Skating Competition," CBS Olympic Coverage, February 21, 1992.
41. "Michael Speaks," *Ebony*, May 1992, p. 40.
42. William Safire, "Dollie and Johnny," *New York Times*, September 7, 1992.
43. Barbara Spector, "The Love of Science: Do Parents Pass It Along to Their Children?" *The Scientist, 5*, September 30, 1991, p. 1.
44. Anthony Lewis, "Politics and Decency," *New York Times*, April 4, 1991, opinion/editorial section.
45. Lawrence Wright, "The Man from Texarkana," *New York Times Magazine*, June 28, 1992.
46. Steven A. Holmes, "For Buchanan Aide, Genetic Conservatism," *New York Times*, February 7, 1992.
47. Alessandra Stanley, "When Ms. Right Falls for (Gasp!) Mr. Left," *New York Times*, April 20, 1992. p. A1.
48. Richard Herrnstein and Charles Murray, *The Bell Curve* (New York: The Free Press, 1994).

MAKING DECISIONS ABOUT SOMEONE ELSE'S OFFSPRING

Geneticists and Reproductive Technology

SIMONE BATEMAN NOVAES

Centre de Recherche Sens, Ethique, Société (CNRS) Paris, France

Recent development in genetics research and the elaboration of a number of new testing techniques have once again triggered debate about the legitimacy of medical practices that aim at detecting and preventing hereditary and/or genetically determined disease.[1] Testing techniques do, in fact, furnish information concerning the genetic makeup of an individual, including the unborn. Used in the context of clinical genetics, they make available to the person (or persons) consulting more precise data about hereditary conditions in the family, permitting that person to make informed decisions for immediate or long-term action, in particular with regard to reproductive choices.

The debate over the clinical use of these techniques usually revolves around certain consequences of obtaining genetic information (for example, the elimination through abortion of individuals who are known to have or are presumed to be affected by the condition) and the risks of such information being misused (for example, discrimination on the job market against persons at risk for certain conditions). They rarely question, however, the very principle on which these clinical practices are based – that individuals concerned by the risk of genetically determined disorders will necessarily find it of interest and of value to them to seek and obtain more precise information. The effects of this principle on the reasoning and the actions of those persons directly involved in these practices (not only clients and their families but also physicians, genetic counselors,[2] and laboratory personnel) are thus for

---

This paper is based on fieldwork completed in 1990, on the organization of semen banking and artificial insemination; among the problems studied were those raised by the genetic aspects of artificial insemination. See Novaes, *Les Passeurs de Gamètes* (Nancy: Presses Universitaires de Nancy, 1994).

Michael Fortun and Everett Mendelsohn (eds.), The Practices of Human Genetics, 169–184
©1999 *Kluwer Academic Publishers. Printed in Great Britain.*

the most part ignored, a practice that hinders a better understanding of the social interaction within which clinical genetics operates and from which many of the fundamental issues arise.

This principle of the desirability (and the rationality) of voluntarily seeking to know more about one's genes also presupposes that the action to be taken in the light of such information will be self-evident to all the protagonists involved. In practice, this is rarely the case, creating a breach in the expected course of action from which its underlying rationale can be examined. This paper, therefore, explores how the principle of seeking genetic information for medical reasons affects decision making in the relationship between the client and the physician in a concrete situation, that of recourse to reproductive technology with donor gametes in France. It studies the dilemmas that the availability of genetic information creates, as conflicts between physicians and clients arise regarding reproductive choices.

Why emphasize reproductive choices? And why use reproductive technology to illustrate this point? Genes imply heredity, thus family antecedents and descendents, and genetic information is most often sought in the medical context by individuals who have reasons to be concerned about "passing on their genes."[3] Reproductive technology in France concerns infertile couples, that is, couples who apparently require medical intervention to conceive. These couples are not necessarily concerned about their own genes, but may be concerned about the health and genetic makeup of a potential donor. In such a clinical situation, the beneficiaries of the practice do not have the means of controlling and carrying out their own reproductive decisions. This reality highlights the difficulties that arise in the medical setting when genetic information is brought to bear on a decision, and it brings into sharper focus the issues involved in seeking and obtaining information about genes.

Genetics and Reproduction

Genetic counseling is one of the oldest medical approaches to the prevention of hereditary disease. By drawing a family tree with the help of information provided by the client(s), the geneticist can calculate the statistical probability that this person, who believes or knows that he or she is a (healthy or sick) carrier of a hereditary condition, will transmit the condition to his or her offspring. Theoretically, it is up to the persons consulting to decide, in the light of such information, whether or not they wish to embark on a pregnancy. However, this does not exclude advice from the genetic counselor, who often suggests to clients what they might do or what "ought to be done" in the light of the information they have obtained.

More recently, as a complement to genetic counseling, different techniques have been devised to determine more precisely whether an individual is a carrier of a genetic condition or whether he or she is at risk of developing such a condition eventually. Certain testing techniques, such as karyotyping and testing for heterozygosity, concern adults with family antecedents of genetic disease, women having had obstetrical accidents (miscarriages, stillborn babies), and adults belonging to a population at risk for a recessive condition (for example, Tay-Sachs disease or sickle-cell anemia). Several techniques, referred to generally as "prenatal diagnosis," may be used to examine the unborn child of adults whose family history indicates the possibility of transmitting an hereditary condition or whose situation includes factors that indicate a risk of discovering a genetic disorder in their offspring (for example, the woman's age). However, in the case of a positive result, prevention implies a late abortion. Preimplantation diagnosis on the embryo would offer the possibility of such testing before the embryo is in the womb and thus would circumvent the trying experience of a late abortion, but such a procedure implies recourse to in vitro fertilization. All of these possibilities widen the spectrum of potential clients for genetic testing and multiply the options they have to obtain more detailed information about their genes.

Reproductive technology – that is, essentially artificial insemination and in vitro fertilization – is not directly concerned with the prevention of genetically determined conditions: in France, it is used essentially to compensate infertility problems encountered by heterosexual couples, for whom an effective treatment cannot be found. However, besides preimplantation diagnosis, the option of turning to a gamete donor in certain cases of infertility introduces the question of prevention, at two levels. First, the donor's general health is usually controlled, so as to avoid inadvertent transmission of disease, including any serious hereditary condition. Second, some couples, who risk transmitting a serious hereditary condition that cannot be detected by the usual prenatal diagnostic techniques, can avoid this risk by resorting to donor gametes. Reproductive technology, usually referred to in France as "medically assisted procreation,"[4] is, therefore, necessarily caught up in the debate over the legitimacy of medical practices that aim at detecting and preventing genetic conditions.

Physicians managing reproductive practices with donor gametes occupy a delicate position with respect to this debate. As physicians responsible for banks mediating anonymous exchanges between gamete donors and persons requesting assisted conception – and in the absence of any type of regulation concerning such transactions – they exercise exclusive decision-making power over the selection criteria used to constitute their stock of gametes, criteria that are often both medical and social. Moreover, they control access to the means – both gametes and instruments – which enable certain couples to conceive children; they can thwart these persons' reproductive projects if they refuse access to these means. Physicians in charge of these reproductive practices are, therefore (with respect to their clients), in an even stronger position than genetic counselors, because they can intervene directly in decisions affecting other people's offspring, on the basis of genetic information they have obtained.

Is the attempt to prevent genetically determined conditions in the context of reproductive technology a legitimate medical activity? What are the particular ethical dilemmas that this raises, and what is socially at stake in such practices?

With respect to genetics, it is customary to reflect on social and ethical issues by making projections on the development of such practices and imagining diverse scenarios of their probable consequences. But I believe that such scenarios are not really useful unless they are based on research leads currently being explored. Projections into the future are all the more perilous in that they cannot take into consideration unforeseen events, which always weigh on the development of scientific practice. I am proposing another approach, therefore, one based on the analysis of a current preventive practice. I would like to show that ethical dilemmas are present at the very heart of what first seem to be purely technical problems, and that it is by looking at how protagonists handle conflict and decisions about these problems that we can best grasp the social consequences of such practices.

The French Federation of CECOS Semen Banks

The French CECOS[5] Federation is a network of 20 independent nonprofit semen banks, usually set up in a public university hospital setting, which function according to a common set of principles when dealing with the problems raised by artificial insemination with donor semen. The first bank was established in 1973 by Georges David, a physician who was concerned with practice standards for artificial insemination in France, as well as with the frequent financial exploitation of distressed infertile patients in the private practice setting. The activity of these banks has grown since and now includes the control of the use of donor gametes for other reproductive techniques and the cryopreservation of embryos awaiting transfer after in vitro fertilization. Their practice was often used as a reference in the French debate concerning the need for legislation on reproductive technology and other new biomedical procedures.

CECOS physicians define their task as mediators in the donor-recipient relationship, who protect *donor anonymity* and *recipient secrecy* regarding recourse to artificial insemination, while guaranteeing the *professional quality of the medical attention* both parties will be receiving. Furthermore, as physicians, they do not wish to be responsible for social innovation in family relationships; thus they strictly limit their practice of artificial insemination to *medically justified situations of infertility*. Therefore, as in the case of most semen banks, CECOS physicians handle donor recruitment and screening but, unlike most banks, they also control access to donor insemination: CECOS frozen semen is available only to heterosexual couples in which the husband has a medically proven infertility problem or an important risk of transmitting a serious hereditary condition. Also, as is the case with most banks, semen donors must agree to remain anonymous; however, they are *not* paid, and only married men (or men established in long-term relationships) who have children and whose wives agree to semen donation are accepted as potential donors. The aim of this policy is to improve the public image and moral standing of donor insemination, thus helping the donors and recipients themselves overcome doubts about the social acceptability of such a transaction. One of the main consequences of this policy, however, has been a scarcity of donors for the banks.

The CECOS Genetics Advisory Board

Clinical genetics was introduced to CECOS activity via two technical preoccupations: screening the semen donor and establishing the validity of certain genetic indications for assisted conception.

Most semen banks in France are concerned with preventing the inadvertent transmission of disease through semen: this concern includes serious hereditary conditions. For this reason, among the exams required of the donor, there are a karyotype (cytogenetic analysis of the chromosomes) and a medical history of the donor and of his antecedents (along with a drawing of the family tree). Time and experience led CECOS physicians to realize that the genetic aspects of screening a donor were more complex than they had first thought. The karyotype did not seem to be a very useful test: very few donors had been excluded because of an anomaly in their karyotype. The interview with the donor to establish his family history seemed, on the other hand, to be the most important part of the control. However, the quality of the information obtained and the accuracy of the decisions made on the basis of this information seemed to depend both on the time spent in interviewing the donor and on the competence in genetics of the person taking the history.[6]

Semen banks also receive requests (about 1% to 2% of overall demand) from couples who risk transmitting to their offspring a serious hereditary condition: one of them may be a carrier of a dominant disease, or both may be carriers of recessive genes responsible for a serious condition. Prenatal diagnosis is an option in some of these cases, but some conditions cannot be detected by existing techniques. Using donor semen can therefore be considered a valid alternative if the couple wishes to have children. However, given constant progress in prenatal diagnostic techniques, the criteria justifying genetic indications for assisted conception with donor gametes need constant reevaluation.

The idea of creating a genetics advisory board was motivated originally by concern about the quality of professional practice: certain decisions required consulting a geneticist, and nongeneticist practitioners involved in screening donors needed training in genetics. At the same time, there was also a practical concern, related to the principles guiding how they practiced: because of the scarcity of donors, it was essential to eliminate as few of them as possible. In fact, a perfectly acceptable donor could be refused on the basis of his family history (and vice versa), if the person controlling the donor's health had no training in genetics. There was also a desire to harmonize practices deriving from the principles to which all CECOS banks (as part of a federated network) adhere.

The Genetics Advisory Board is composed of three types of members: (1) CECOS physicians (clinicians and biologists), who submit the cases and ask the questions; (2) young clinical geneticists, working with the CECOS banks, most of whom have training in pediatrics; and (3) clinical geneticists and cytogeneticists, who do not work in CECOS banks but who are invited as consultants because of their specialty. Its initial task was to elaborate a set of guidelines and practical recommendations; but as CECOS physicians tested these guidelines by confronting the board with specific cases from their daily practice, the board came to realize that the setting of practice standards could not be handled from a purely theoretical perspective, as it "necessarily entailed reflection on the very nature of artificial procreation with donor gametes, on its meaning, its justification, its risks, and the limits that should be assigned to it."[7] The board subsequently changed its mode of functioning and adopted a new approach based on case histories: difficult or unusual cases are now submitted by the 20 banks to the board for an opinion; the board recommends a course of action and, at the same time, when necessary, revises its own guidelines.

At its first meeting in 1983, the board established a list of genetic factors justifying the exclusion of a donor candidate (essentially, major chromosomal alterations, dominant disabling pathology, and some frequent, serious recessive conditions that might be found readily among recipients). In fact, three rather than two categories of donor candidates were created: (1) those excluded because of a serious risk of transmitting a genetically determined condition; (2) those accepted without reservation (there is no apparent risk of transmitting a major handicap); and (3) those accepted with a *cumulative risk factor* (CRF), that is, donors whose family history indicates a risk of transmitting a genetically determined condition, which, however, is either not serious or very rare, and can be prevented by attributing that donor's semen to recipients whose history does not present the same factor.[8] The creation of this last category of donors, which allows for fewer exclusions, was an attempt to avoid the slippery slope toward excessive rejection in the screening of donors and, at the same time, a response to concern about the unfounded exclusion of the scarce benevolent donor that the CECOS banks were seeking. However, this new category requires at least limited screening of the recipient. The problems raised by selection, which all genetic screening implies, are thus displaced from control of what semen will be kept for cryopreservation to decisions on how semen will be assigned for insemination.

This new genetic emphasis on selecting semen for a particular recipient (which is now added to criteria concerning the morphological characteristics of the couple – usually skin color, blood group, and sometimes hair and eye color) highlights the physician's role and responsibility as a mediator in momentarily associating for reproductive purposes two people who are not to be recognized socially as the child's parents. The donor remains anonymous, and recipients usually prefer to keep recourse to artificial insemination secret. The criteria and purpose of such matching must therefore be made clear to all the protagonists involved – and, ultimately, to society, if donor insemination is to be considered an acceptable procedure.

Three Ethical Dilemmas

Among the different cases examined at each meeting, there are always borderline cases, which question not only the commission's guidelines but also the apparent clarity and self-evident nature of CECOS principles regarding "assisted conception." Moreover, clinical geneticists working in this area are constantly forced to reconsider the ethical assumptions on which their preventive practices are based, as they encounter novel situations in which existing standards and normative references do not seem to apply.

Following are three recurrent dilemmas encountered at these meetings; they enable us to identify and characterize new decision-making problems – perhaps a step toward understanding how to solve them. Their resolution does not seem to involve finding a right answer but, rather, reaching an agreement among all concerned about who will make certain decisions, on what basis these decisions should be made, on whom responsibility rests for their consequences, and if such responsibility is to be shared, by whom.

1. At what moment is the transition from sexuality to medically assisted conception justified?

This dilemma arises when the validity of certain genetic indications for assisted conception is being considered. Contrary to what might be expected, the geneticists on the Genetics Advisory Board are quite reticent about pushing a couple into a situation in which they must abandon the idea of conceiving on their own. The couple is not infertile and, in the case of recessive conditions, each partner would have been able to conceive children without that particular risk had they been with another partner, a noncarrier of the same trait. Moreover, recourse to artificial insemination with donor semen requires the male partner to use some form of contraception during the period of insemination. Abstaining from using one's procreative capacities is often difficult to accept, even when one is perfectly conscious of the risks involved. In fact, in certain instances, the child born after insemination was afflicted with the disease that recourse to donor gametes should have prevented, just because no contraception had been used. When the male partner has a dominant condition, the problem is simpler to resolve: he can decide to have a vasectomy, which permanently avoids the risk of transmitting his condition. Donor insemination is then more easily justified on the grounds of infertility.

In these cases, geneticists on the board often prefer recommending genetic counseling and prenatal diagnostic techniques. They propose donor insemination as an alternative only when the genetic condition under consideration is *serious*, when the risk of transmission is very *high*, and in such cases where no genetic testing for the condition is available.

It is thus the high probability of conceiving a child seriously stricken by a genetic condition (or repeated unsuccessful attempts to conceive a healthy child, despite a favorable probability of doing so) that seems to justify the move to medically assisted conception. In other words, the couple's sexual relations must be perceived by physicians *as being, for all practical purposes, sterile*[9] before access to assisted conception is viewed as medically justified. This prudent attitude nevertheless implies a "tentative pregnancy,"[10] with the risk of a late abortion in the case of a positive diagnosis. A certain number of requests for donor insemination on the basis of genetic indications have come from couples who have had to endure the experience of late abortion two or three times, despite the low probability of conceiving a child with a particular genetic condition. It is very difficult to be "reasonable" by adopting an attitude based on risk calculation, when chance seems systematically to come out against the birth of a healthy child. Moreover, according to some geneticists, couples facing this trying situation are not adequately accompanied through the ordeal of a positive diagnosis by the hospital personnel in charge of prenatal diagnostic techniques.

Couples requesting donor insemination for genetic reasons have been therefore dissuaded from resorting to assisted conception before they have exhausted all the possibilities of conceiving a child on their own (in other words, before they have been considered effectively sterile). This however does not spare them the experience, in particular the woman, of a highly medicalized pregnancy.

2. To what extent is a physician's responsibility involved when, in the context of assisted conception, there is a risk of transmitting an hereditary condition?

Eliminating all risk of transmitting a genetically determined condition to one's offspring is an illusory objective, given the fact that every individual is most probably a healthy carrier of several recessive genes for serious conditions. Therefore, in the context of recourse to reproductive technology, the question becomes: what professional attitudes can be defined as *acceptable risk taking*?

This question is raised diversely in three different types of situations: genetic screening of the donor; the eventual screening for a genetic condition of the woman to be inseminated; and possible recourse to prenatal diagnosis in the case of a pregnancy after donor insemination (or donor in vitro fertilization).

The purpose of donor screening is to avoid transmitting by negligence, among other things, serious genetic conditions. CECOS physicians feel that it is unacceptable to use the semen of a man determined to be (or suspected to be) a carrier of a *serious* genetic condition, which – in the case of certain recessive conditions – can be *found frequently* in the population, and for which there is a *high* risk of transmission. When the genetic origin of a condition has not been clearly established (for example, in the case of a donor who has been cured of cancer), they also feel that, in the absence of conclusive data, the physician must not take the responsibility of accepting such a donor. Risk taking in this context concerns someone else's offspring.

These general attitudes are far from being clearly defined standards of practice, however. Even though there is no attempt here to eliminate "bad" genes (in order words, no systematic negative eugenics), the interpretation of terms such as "serious," "frequent," and "high" poses some problems. For example, is hereditary deafness, some forms of which can be operated on (otospongiosis), a serious condition? Even though there may be agreement among board members with respect to the objective data defining a particular condition (its characteristics and the calculation of the probability of transmission to one's descendants), that is often not the case when it comes to deciding what action should be taken. Such a decision implies subjective evaluation of medical data and statistical calculations, and physicians refer to their own experiences and values in the areas of procreation, health, and handicap, to help them make the difficult choices – values that are not necessarily professional, moreover. It is therefore difficult to define consensually the concrete limits of acceptable risk taking.

On the other hand, these attitudes are sometimes in contradiction with CECOS principles, which affirm – rather simplistically from a genetic point of view – that only men in good health with children in good health are accepted as semen donors. A couple requesting donor insemination enters into it with an idea of a donor in good health, which may refer to normative criteria quite different from that used by physicians to select their donors. It is important, therefore, that couples requesting assisted conception be informed of the criteria underlying physicians' decisions regarding what constitutes an acceptable risk.

But is a physician responsible in the same way for the outcome of a pregnancy by donor insemination, when the risk of transmitting a genetically determined condition proceeds from the inseminated woman's antecedents? In some cases, physicians discover that the woman, whose partner is infertile, is herself a carrier of a serious dominant condition, but she may have it in only a minor form (which also may be why it was overlooked at first). In other instances, she might be a carrier of a serious dominant condition (such as polyposis of the colon), which is a late-onset disease of variable penetrance. Contrary to the situation in which it is decided to exclude a potential donor – a decision that seems justified because it is someone else's offspring who will be affected – the decision to refuse donor insemination on the basis of the woman's family history (even when this attitude seems medically justified) is nevertheless recorded by board geneticists as "questionable."

Does a physician have a right to refuse assisted conception to an infertile couple on genetic grounds? Is there a dividing line, and if so, where is it, between the physician's responsibility for controlling the medical factors that intervene during a pregnancy and the couple's autonomy in making decisions about their own reproductive lives? Is the fact that the birth of a handicapped child implies expensive medical treatment (usually paid for by the state) a valid counterargument against the fact that ultimately it is the parents who assume responsibility for raising the child? In other words, who assumes responsibility for the consequences of risk taking in childbearing, and how does this affect their right to decide?

The same problem arises when, given the woman's antecedents, pre-natal diagnostic techniques are mentioned as an alternative, permitting surveillance of her pregnancy after donor insemination. In such cases, CECOS physicians almost always reject this option, because they feel that aborting a pregnancy that has been obtained with scarce donor semen is unacceptable. This attitude contrasts notably with their insistence that couples who request donor insemination, because of the man's genetic antecedents, first attempt prenatal diagnosis after conceiving on their own – despite the risk of late abortion. So, in the case of assisted conception, the question becomes: who in fact assumes responsibility for the risk of a late abortion?

Medical justification for refusing donor insemination (or any other techniques implying gamete donation) rests on the argument that what may be an acceptable risk when it is entirely in the couple's hands is not necessarily acceptable in the context of assisted conception. Two elements distinguish this context from what physicians call "natural" conception: recourse to a donor and recourse to a physician. CECOS physicians often feel that, as trustees of the donor's semen, they are accountable to the donor for the way in which they utilize the semen. The donor might not approve of his semen being used in cases that involve considerable risk of the child's being born with a serious hereditary condition for example. Moreover, as physicians, they cannot abstain from evaluating such situations according to professional criteria, and assuming consequent responsibility for actions based on professional judgment. A third argument, used more sparingly, is that of the "child's interests"; this argument seems to underlie the other two, however. For in fact, each time the physician's professional and moral responsibility with regard to donors is mentioned during the discussion of a particular case, what is usually at stake is a possible miscarriage, a late abortion, or a stillborn infant. The preceding arguments reveal, nevertheless, that these novel reproductive situations are not clearly understood, in terms of the respective rights and obligations of each of the protagonists, and that they ultimately bring to the fore a question that is impossible to answer: who can best represent and defend the interests of the unborn child?

The notion of medical responsibility, as it is used in board discussions, is strongly related to the fact that CECOS physicians believe they should dispose of semen (or ova) entrusted to them for the purpose of helping an infertile couple conceive a healthy child. If the birth of a healthy child seems unlikely, CECOS physicians feel they must abstain from using donor gametes. This construction of the notion of medical responsibility defines the physician's role, because of the control he exerts over donor gametes, not as a simple mediator between donors and recipients but as a major protagonist in reproductive decision making. In fact, if the physician feels responsible for the outcome of such a pregnancy, we can hypothesize that the act of assisted conception is a different medical act from others; it is a reproductive act, and by thus assisting an infertile couple to conceive, the physician is directly contributing to the birth of a (healthy or afflicted) child. This idea of the physician's responsibility, emanating from practical experience, contrasts with the usual formulation of medico-legal responsibility in France: usually physicians are required to put all possible therapeutic means at the patients' disposal, but are not held responsible for the ultimate outcomes the procedure.[11]

3. Should a physician use gametes for donor insemination that do not endanger the child's health or life, but that do physically mark the child with his or her unknown genetic origins?

Some donors have hereditary traits that could indicate that the parents sought donor insemination; in other words, the traits make it obvious that the father did not conceive the child. This is a problem because most semen banks in France guarantee that the choice of semen for donor insemination will make it plausible that the child was conceived by his parents.

Several genetic traits fall into this category, but each one poses different problems. A donor, who is a known heterozygote for a recessive condition that neither of the parents has, has a 1/2 possibility of transmitting this trait to the child. However, the trait is not visible, so at most the child would be a heterozygote for this condition too, and therefore healthy; this trait would not reveal publicly the parents' recourse to donor insemination. In certain cases, a cleft palate and a harelip are hereditary malformations; although they are visible, they are operable defects, and in any case, do not necessarily function as markers, as sporadic (nongenetically determined) cases are also possible. On the other hand, a polydactyl donor (having more than the usual number of digits on hands or feet) has a visible hereditary trait, which does not threaten the child's health but which acts indisputably as a marker of donor insemination.

Accepting such a donor presupposes that couples requesting donor insemination are aware of how the genetic screening of donors is conducted and of the criteria physicians use to guide them in selecting donors in good health. The acceptance of candidates with visible hereditary traits that have no consequences on the child's health presupposes that couples are ready to surpass the need for secrecy, since the marker transforms recourse to donor insemination into a visible, physical fact. However, because secrecy is the standard concerning donor insemination, there is reason to believe that this means of conceiving children and the kinship it establishes is not always perceived and experienced as legitimate.

Acceptance of such donors also implies that geneticists redefine their clinical objectives in the context of assisted conception. Should they try to prevent hereditary diseases by avoiding the deliberate transmission of harmful genes; or should they only try to prevent the most disabling hereditary conditions by placing primary emphasis when making decisions on the seriousness of a condition? If they choose the latter, it must not be forgotten that evaluating the seriousness of a condition is never totally objective. Moreover, the way this task is defined necessarily increases the physician's involvement in what his or her clients perceive as a desirable family life.

Conclusion

An increase in medicine's capacity to detect and eventually to prevent genetically determined conditions and malformations does not necessarily mean an increase in control over genetic parameters: they are too numerous, and the relationships among them too complex to be totally under control, even in a single, specific situation. Conceiving and childbearing, whatever the conditions, always involve a certain degree of risk taking, and a physician intervening in the context of reproductive medicine cannot in any way guarantee the birth of a perfectly healthy child. At most, physicians can attempt to maintain risk taking at a level defined as acceptable by their profession.

In the apparently technical endeavor to define acceptable limits to risk taking, ethical dilemmas arise, as protagonists are confronted with the possibility of novel behavior or innovative action, which appears questionable in the light of usual normative references. In the concrete cases that I have analyzed, the situation can be summarized as follows:

1. The real or effective sterility of a couple's sexual relations, when these do not lead the couple to abandon their desire to conceive children, justifies assisted conception. Such medical intervention cannot be strictly defined as therapeutic, however, because it does not cure one or both partners' infertility. Medically assisted conception creates a new relationship between a physician and his or her (otherwise healthy) patients, situated in an institutional "no-man's land," somewhere between the family and medicine, in which the appropriate normative references for decision making are not clear.
2. Preoccupation with the genetic dimensions of assisted conception leads physicians to define their responsibility in a way that admits their direct intervention in reproductive decisions – because of the fact that physicians in this context are contributing directly by their action to the birth of a child.
3. This construction of medical responsibility encounters obstacles nevertheless, in the form of reservations or objections on the part of patients or other physicians, which reveal that the limits that might be imposed on such responsibility cannot be argued solely on the basis of medical criteria.

In fact, every medical decision implies a subjective evaluation of the situation in which treatment is required: the choices made by physicians are informed by professional experience, which varies from one physician to another according to sociobiographical elements. Even though social and moral values related to "natural" reproduction are perceived by physicians as a foreign and illegitimate basis for their decisions, when the action to be taken in a particular case does not seem self-evident, those values will regain importance.

Defining an acceptable line of medical conduct with regard to the detection and prevention of serious hereditary conditions cannot be restricted to an area defined as "strictly medical" or "strictly scientific," therefore. At the very heart of what appear first to be purely technical questions are a series of choices that affect society regarding the best way to conceive children. The situations bring into play our ideas concerning normality, health, and illness; our notions about responsibility toward the unborn, including questions about who has the responsibility to make decisions about their welfare and future; and the legitimacy of kinship established through medical intervention. For this reason, the criteria guiding medical choices in detecting and preventing genetically determined conditions need to be aired and discussed publicly. Moreover, once the limits of such practices are set, they can be considered morally and socially acceptable only – if even after wide social approval – they remain open continually to critical review and reevaluation.

Acknowledgment

I thank Gwen Terrenoire for pertinent comments and suggestions on the first draft of this paper.

Notes and References

1. As we will see later in this paper, both detection and prevention are problematic notions in genetics. Detection may sometimes only mean establishing a statistical probability that a condition will or will not occur. Prevention may simply entail abstinence from procreating or selective abortion, as there are as yet no successful remedies for many genetically determined conditions. Fundamentally, prevention involves making choices about risk taking, on the basis of a statistical probability.
2. It must be noted that the distinction between physicians and genetic counselors is not pertinent in France, where genetic counseling is done exclusively by physicians.
3. This familial dimension of genetics is often absent in studies that deal with the more technical aspects of laboratory work in molecular biology.

4. In French, I usually distinguish procreation, which is a social process, from reproduction, which is a biological process. Procreation is, in other words, the social organization of biological reproduction. Thus the term "medically assisted procreation" translates quite well the idea of a *socially* organized solution to the absence of descendants. Nevertheless, I have chosen, in this article, to respect the terminology familiar to English-speaking readers: reproductive technology and reproductive choices.

5. Centre d'Étude et de Conservation des Oeufs et du Sperme Humain, in other words, Center for the Study and Preservation of Human Eggs and Sperm. For an extended sociological analysis of the ethical, institutional, and technical dimensions of semen banking and donor insemination in France, see Simone Novaes, *Les Passeurs de Gamètes, op. cit.*

6. J. Selva et al., "Genetic Screening for Artificial Insemination by Donor (AID): Results of a Study on 676 Semen Donors," *Clinical Genetics, 29* (1986), pp. 389–396.

7. P. Jalbert and G. David, "Problèmes génétiques liés à la procréation artificielle par don de gamètes: solutions adoptées par les CECOS," *Journal de gynécologie obstétrique et de biologie de la reproduction, 16* (1987), p. 548.

8. The most common examples of cumulative risk factors are allergies and cardiovascular disease. For more detail on these guidelines for the genetic screening of donors, see P. Jalbert *et al.*, "Genetic Aspects of Artificial Insemination with Donor Semen: the French CECOS Federation Guidelines," *American Journal of Medical Genetics, 33* (1989) pp. 69–275.

9. I believe that a definition of infertility, not just as an incapacity to conceive (and bear) a child but as an incapacity to produce a *healthy* child, will play a major role in the way both reproductive technology and genetic diagnosis and therapy will evolve as medical practices in the future.

10. Here I am using the term coined by Barbara K. Rothman in her book *The Tentative Pregnancy* (New York: Viking, 1986), an excellent analysis of prenatal diagnosis, both as a psychologically trying experience and as a practice that has consequences on the social attitudes toward handicaps. In her conclusion, she recommends that prenatal diagnosis be replaced more frequently by donor insemination because it is less traumatic for the woman and, as a practice, would have fewer long-term consequences in terms of discrimination. She seems to disregard the complexity of the problems raised by donor insemination, however, which, as we will see, lead in a different manner to similar consequences in social attitudes.

11. In French, the physician has an "obligation de moyens" but not an "obligation de résultat."

PKU SCREENING

Competing Agendas, Converging Stories

DIANE B. PAUL

University of Massachusetts at Boston

In 1963, Massachusetts became the first state to initiate mandatory genetic screening of newborns for phenylketonuria (PKU), an autosomal recessive disorder whose incidence in the United States, Britain, and most of Western Europe is between 1 in 11,000 and 1 in 15,000 births.[1] Although aspects of the pathogenesis and population genetics of PKU remain obscure, it has been known since the 1950s that the disease results from a defect in the enzyme phenylalanine hydroxylase, which catalyzes the conversion of phenylalanine (an essential amino acid found in most foods) to tyrosine. In the absence of therapy, phenylalanine accumulates to disastrous levels in the blood. The consequences include severe behavior problems and mental retardation. About 90% of those affected have IQs of less than 50.[2]

However, the disease is treatable by a diet restricted to special phenylalanine-free foods, supplemented by a formula with the other essential amino acids and extra tyrosine. In 1946 Lionel Penrose suggested that its effects might be alleviated "in a manner analogous to the way in which a child with club-feet may be helped to walk."[3] Within a few years, studies suggested that dietary treatment would in fact bring some improvement, and a number of screening programs were initiated. In 1961, Robert Guthrie invented a cheap and simple blood test suitable for mass screening.[4] The conjunction of the Guthrie test with a statistical investigation validating earlier studies strengthened the existing movement to screen newborns for the disease.[5] Massachusetts quickly instituted a large-scale pilot program utilizing the Guthrie test. Within the year, and without any legislative requirement, every maternity hospital in the state was screening all newborns for PKU.[6]

But hospitals in many other states were slow to begin screening, and pressure mounted for legislated programs. By 1975, 43 states had passed screening laws, and 90% of all newborns were being tested.[7] (None

185

Michael Fortun and Everett Mendelsohn (eds.), The Practices of Human Genetics, 185–195
©1999 *Kluwer Academic Publishers. Printed in Great Britain.*

mandated treatment).[8] The laws were passed despite considerable opposition from the medical community[9] and researchers in the field of human metabolism. While private practitioners resented state interference with the doctor-patient relationship and feared an increase in malpractice suits, the researchers questioned the reliability of the Guthrie test and denied that enough was known about the prognosis and management of the disease to justify mandatory screening. However, the critics proved no match for PKU clinicians and lay organizations, in particular the National Association for Retarded Children (NARC), and its allies in the Children's Bureau of the Department of Health, Education, and Welfare and the state health departments.[10] Today every U.S. state screens newborns for PKU, and usually other metabolic disorders as well.[11]

Why did such a rare condition, affecting only about 400 American infants a year, generate such intense activity on the part of clinicians, parents, and public officials? Paul Edelson has noted the disparity between the modest impact of PKU on the nation's health and the massive campaign pursued to control it.[12] He has also suggested a number of explanations: for scientists, PKU research made Archibald Garrod's biochemical genetics suddenly relevant to clinical medicine; for clinicians working in institutions for the retarded, it moved the study of mental deficiency into the sphere of modern scientific medicine; most important, for parents and public officials, it provided an example of a form of retardation that was treatable, indeed, that would allow affected individuals to lead normal lives. "In the face of such dramatic possibilities," he argues, "arguments regarding a lack of placebo-controlled studies, or the suggestion that PKU laws intruded on the doctor-patient relationship had little impact on legislators."[13]

Mass screening was also extended to some countries unable to provide even a minimum of medical services. Through the sale of surplus farm commodities under Public Law 480 (the Agricultural Trade Development and Assistance Act of 1954), the U.S. acquired foreign currencies that were made available to federal agencies for medical and scientific research and other health-related activities abroad. These funds were used to answer a question of interest to the Children's Bureau: to what extent does the incidence of PKU vary with race and ethnicity?

Because PKU is such a rare disease, some early screening programs identified few if any cases. Officials in Washington, D.C. reasoned that they had better things to do with their money and ended the program; some states threatened to follow suit. It seemed that the efficiency of screening might be increased if particular populations were targeted. But in the mid-1960s, only the general population incidence of PKU was known. In hopes of obtaining data on different racial and ethnic groups, Children's Bureau staff proposed initiating screening programs in some "PL-480" countries.[14]

They did not anticipate a positive response given the extent of unmet basic maternal and child health needs in the East European and Third World countries where most PL-480 funds were available and also the obstacles to providing treatment when cases were found. Children's Bureau analysts were thus surprised when researchers in some countries, including Poland, Yugoslavia, and Pakistan, proved enthusiastic. Researchers' eagerness to cooperate, they suggested, arose "partly from their desire to be associated with West in something that is new and exciting, and partly from their realization that this program gives them an opportunity to develop laboratory and clinical facilities that can be used for a much broader program than the detection of one, rare inborn error of metabolism."[15] Even in countries with more pressing health problems, they found, "there is a strong desire to work on special problems, which gives the workers a sense of belonging to a modern scientific community. We believe the isolation of scientists in some countries, their feeling of being passed over by the march of science, should be taken into account when we determine priorities in our cooperative programs."[16]

In the United States and many other countries, mass screening, combined with early dietary treatment, did succeed in eliminating PKU as an important cause of mental retardation. For this reason, newborn screening for PKU is often cited as the premier success story of applied human genetics and as proof of the value of genetic medicine. PKU is "the paradigm therapeutic case" of postnatal diagnosis;[17] routinely characterized as "a genetic screening success story;"[18] and a "model of successful intervention in the prevention of mental retardation."[19] The geneticist Charles Scriver writes that therapy for PKU "became not only an epitome of the application of human biochemical genetics, but also a model for so-called genetic medicine and for public health."[20] PKU is also invoked as a precedent by those who wish to expand screening programs. Thus after noting the low incidence of PKU, *Wall Street Journal* reporters Jerry Bishop and Michael Waldholz note, "If mass screening can be justified for...relatively rare genetic disorders, then screening newborns for susceptibility to such common diseases as diabetes, schizophrenia, coronary heart disease, or cancer would seem even more worthwhile."[21]

But the PKU story is more complex than these (typical) accounts suggest. Mass screening has indeed prevented retardation in tens of thousands of individuals. There is no doubt that the vast majority of

affected infants are better off with treatment than they would have been
in its absence. But PKU screening is not an unqualified success. Several
problems were recognized in the early 1960s and explain some of the
initial opposition to mandated programs (although physicians' resistance
to any state dictates and to concerns about malpractice suits were also
significant factors).[22] Others were only recognized after years of experience. But as the realities have become increasingly complex, the stories
have become increasingly simple. What do we know of the compli-
cations? And what explains their near invisibility outside the professional
literature?

In the 1960s, opponents of mandated screening expressed concerns
about the many gaps in the medical understanding of PKU, the validity
of the Guthrie test, and the efficacy of the recommended dietary
treatment. They argued that mandated screening was premature and
likely to result in a reduced commitment to research.[23] In 1975, the
Committee for the study of Inborn Errors of Metabolism of the National
Research Council admitted in a generally positive report that "screening
was started, frequently under mandatory laws, when questions regarding
diagnosis, prognosis, and optimal management were unanswered."[24] At
the time, no one knew what proportion of infants with elevated phenylalanine levels were at risk for retardation, what level of blood
phenylalanine was optimal, whether restriction of phenylalanine in early
life would prevent retardation in infants with PKU, or whether the dietary
therapy could be discontinued after brain growth was complete.[25]

The early years of newborn screening were marked by high false
negative and very high false positive rates, as well as unreliable
laboratory work. A 1970 survey found that for every PKU infant, 19 who
did not have the disease received an initial positive screening test –
reflecting the (then unknown) fact that elevated blood phenylalanine
levels may result from the relatively benign condition hyperpheny-
lalaninemia, as well as PKU.[26] As a result, some infants who did not have
the disease were treated for it, with damaging results.[27] (Today about one
in every 70 affected infants is missed, while the false-positive rate is
about 1%).[28] In 1979, Mark Lappe published a scathing critique,
concluding that, "On balance, PKU screening has probably helped fewer
persons at risk for mental retardation than would have been helped had a
similar effort been made in a preventive health screening program that
measured hypothyroidism."[29]

Most of the initial problems were eventually resolved. But new ones
appeared. The NRC committee noted that at the time screening became
widespread, subjects had not been followed long enough to determine the

extent to which therapy would prevent retardation, or whether specific behavioral or cognitive problems would develop. "Yet most health professionals hailed the diet as highly effective, and there was little organized effort to determine whether, in the long run, screening would meet its objective. Only after a few lonely but loud critical voices were raised" was an effort made to determine optimal phenylalanine levels and to measure the diet's effectiveness.[30]

These studies indicated that initial assumptions about the required length, effectiveness, ease of management, and psychosocial effects of therapy were much too sanguine. Early-treated patients with PKU, while not mentally retarded, generally have lower IQs than normally expected.[31] They do poorly in arithmetic.[32] They often experience psychological problems[33] and reduced visual perception and visual-motor skills.[34] Moreover, these patients are now advised to remain on an expensive and unappealing diet for much longer than was initially anticipated.

In the 1960s, it had been assumed that only the developing brain was vulnerable to damage.[35] But studies in the following decade revealed that IQ scores declined after removal of dietary controls.[36] Children had initially been taken off the diet as early as four years old; most stopped at six.[37] Although there was (and is) no consensus on when the diet should be discontinued or relaxed, recommendations have become consistently more cautious. Researchers and clinicians now generally advise diet continuation through adolescence, and some advocate lifelong restrictions. A recent national survey of treatment centers indicates that 61% of programs now recommend indefinite continuation of the diet for males; 77% recommend this policy for females. (In the previous decade, only 23% of programs recommended indefinite continuation for males, and 42% for females).[38] However, compliance is hard to obtain. The diet is boring and the formula unpalatable. It is difficult to get children and especially adolescents – who want to eat what their friends do – to remain on the diet. It is even harder to get those who have stopped to restart it.

In recent years, the problem that has received the most attention is maternal PKU. If women do not resume the diet prior to conception and maintain it throughout pregnancy, the effects on their offspring are catastrophic, including mental retardation, microecephaly, and heart defects.[39] Susan Waisbren notes that, before the advent of newborn screening, women with PKU bore very few children. Thus screening has converted a rare occurrence into a major problem. Moreover, it is not easy to locate the at-risk adolescent girls and young women. While a few are seen regularly in PKU clinics, most discontinued the diet during childhood and have not been followed for many years.[40] About 2,700 women with PKU will be of childbearing age in the next 20 years.[41] In the absence of a remedy, all the beneficial results of screening may be neutralized by the birth of retarded children to women who have ended the diet.[42]

The issues receiving the least attention are economic. Both the formula and special foods are expensive. The cost to the *pharmacist* of a year's supply of adult formula is about $4,600. The special foods are also costly; for example, a 9-ounce can of white bread costs $3.55.[43] The flour required to make a loaf of low-protein bread runs about $6.00.[44] Less than nine ounces of spaghetti costs $3.35.[45] (There is no U.S. maker of low-protein pasta, which is therefore imported from Europe.) These figures do not include charges for shipping and handling. Who generally pays for the dietary therapy? The states? Insurers? Individuals? What is the practical experience of those families that receive a positive screening result?

Research directed to answering these questions has been meager indeed. We have very little general knowledge about how screening "programs actually feel to those they touch."[46] We know almost nothing at all about how they touch people economically. In its 1975 report, the NRC committee noted that 25 states provided for treatment. Of these, regulations in seven specified that it be free; in one, treatment was to be provided without charge if the doctor requested it; in ten others, if the family were "in need."[45] According to members of the committee, "If all infants are to be screened, then there is an obligation to ensure that all infants discovered to have PKU receive optimal therapy. Adequate means of financing the costs of special diets and other aspects of care for families not covered by insurance and unable to pay must be a societal responsibility."[48] But almost no one has tried to determine how the diet is actually financed, and with what results.

There exists a huge literature on scientific and medical aspects of PKU. There are also many analyses of adolescents and their parents' disease and diet-specific knowledge, the psychosocial and cognitive effects of therapy, and the impact of the disease and diet restrictions on family functioning. A Medline search stretching back to 1966 generated almost 3,000 articles primarily concerned with PKU. Only one discussed the economic impact on families.[49]

Its findings were not reassuring. A survey of three treatment centers conducted by the New York State Department of Health found that most patients who had health insurance or Medicaid coverage were unable to obtain reimbursement for the formula or special foods. Payment was denied to 44% of those with health insurance policies and was covered by only 10% of those eligible for Medicaid. A public program paid food costs for children in upstate New York but not in New York City (where only infants are covered). No financial assistance was available to adults who were ineligible for Medicaid and lacked private insurance coverage for the special foods. Many families found the costs of the special diet to be onerous. The centers' staffs "interceded for patients by appealing to private insurance carriers and to local Medicaid offices to attempt to reverse decisions that had denied reimbursement for special foods. They reported that their efforts were rarely effective."[50] (Of course the inadequacy of financial support for families is typical of many chronic diseases in the U.S.).

If the PKU story is so complex, why is it often described as an unqualified success? The answer lies in the moral lessons the story is employed to teach. Both enthusiasts for genetic medicine and critics of genetic determinism have come to find the story useful. These convergent interests mean that no one has an incentive to pick up the rock and see what lies underneath.

It is obvious why advocates of genetic medicine in general, and screening programs in particular, might be inclined to a cheerful interpretation. But skeptics of testing and screening abound, particularly on the political left. They would appear to have every interest in exposing problems in this model" case. They are surely sensitive to the fact that the existence of an effective therapy, "does not mean that it is actually accessible to the children who need it."[51] They understand that accounts of genetic medicine tend to overestimate the benefits and obscure the costs, that on-the-ground experience often deviates from theory, and that research interests may be advanced in the guise of therapeutic programs.[52] Yet they accept self-serving and wholly abstract accounts of newborn screening. Why?

Critics of screening have their own interest in presenting PKU therapy as an unqualified success. It is a common cultural assumption that what is genetic is fixed. PKU seems to provide a dramatic example of the falsity of that assumption. Although it is an "inborn error of metabolism," a knowledge of its biochemistry enables us to limit the supply of the damaging substrate. To put it another way, PKU is a trait with a heritability of 1.0. But its expression can be drastically altered by a change in environment. PKU thus demonstrates that biology is not destiny. Joseph Alper and Marvin Natowicz note that: "There is a tendency among the lay public to believe that genetic means unchangeable. This belief is false. For example, the invariably serious neurological effects of phenylketonuria...can be largely prevented by providing the affected newborn with a phenylalanine-restricted diet."[53]

PKU screening was transformed into a simple success story during the
1970s, when it became a weapon in the controversy over the genetics of
intelligence. The efficacy of treatment provided a dramatic, decisive, and
easily understood rejoinder to the argument that a high heritability of IQ
rendered futile efforts to boost scholastic performance.[54] The sociobiology controversy served to reinforce this trend. PKU became the
standard example of the flaws of genetic determinism. Philip Kitcher's
discussion is typical:

There is an allele that, on a common genetic background, makes a critical
difference to the development of the infant in the normal environments
encountered by our species. Fortunately, we can modify the environments...and
infants can grow to full health and physical vigor if they are kept on a diet that
does not contain this amino acid. So it is true that there is a 'gene for PKU.'
Happily, it is false that the developmental pattern associated with this gene in
typical environments is unalterable by changing the environment.[55]

This is true – but misleading. The reader would not suspect that the
dietary regime is arduous and that even those adolescents and adults who
fully comply with it usually suffer some degree of behavioral and
cognitive impairment.

It is even more surprising to read similar accounts by critics specifically concerned with screening programs. In *Dangerous Diagnostics*,
Dorothy Nelkin and Laurence Tancredi criticize the "increasing
preoccupation" with tests in American society.[56] They suggest in
particular that testing may enhance "institutional control at the cost of
individual rights."[57] PKU screening might seem an excellent illustration
of their fears. Most states provide for parental objection to screening on
some (usually religious) grounds. In practice, these statutes or
regulations have turned out to be meaningless. Few states require that
parents be told they have the right to object or even that they be informed
of the test.[58] They have no effective right to refuse participation. Yet the
authors use the PKU case only to illustrate how a disease may be easily
controlled by changing the environment. "Though sensitivity to phenylalanine is inherited," they write, "its principal manifestation, mental
retardation, depends on diet. Removing phenylalanine from the diet of
afflicted individuals will avoid the serious retardation that characterizes
the disease. One can, in fact, have the gene, yet with proper dietary
changes never show the manifestations..."[59] The moral: even accurate
detection of a gene will not necessarily eliminate uncertainty about
disability. Ruth Hubbard and Elijah Wald provide an equally upbeat

account of treatment for PKU in support of the (reasonable) claim that treating symptoms is preferable to correcting mutant genes.[60]

The PKU story is infinitely plastic, employed by both celebrants and skeptics of genetic medicine. But it does not serve all interests equally. In his trenchant review of *The Bell Curve*, Robert Wright notes that Herrnstein and Murray are wrong to conclude that "equalizing environments will have no effect" on intellectual performance for "it turns out that if you put all infants on a diet low in the amino acid phenylalanine, the disease disappears."[61] Critics of genetic determinism need not counter one myth with another. There are many ways to demonstrate why "genetic" should not be equated with "fixed." But there are few genetic screening programs that lend themselves to long-term evaluation. The history of PKU screening can teach us a great deal about the social and economic realities of genetic medicine – if we let it.

Notes and References

1. Editorial, *The Lancet*, May 25, 1991 p. 1256.
2. Katherine L. Acuff and Ruth R. Faden, "A History of Prenatal and Newborn Screening Programs: Lessons for the Future," in Ruth Faden et al., *AIDS, Women, and the Next Generation: Towards a Morally Acceptable Public Policy for HIV Testing of Pregnant Women and Newborns* (New York: Oxford University Press, 1991), p. 64.
3. Lionel S. Penrose, "Phenylketonuria: A Problem in Eugenics," *The Lancet*, June 29, 1946, p. 951. This article is a reprint of Penrose's inaugural lecture at University College, London. Penrose actually experimented with a low-phenylalanine diet in the 1930s; See Daniel J. Kevles, *In the Name of Eugenics* (New York: Knopf, 1985), pp. 177–178.
4. Robert Guthrie, letter, "Blood Screening for Phenylketonuria," *Journal of the American Medical Association, 178* (1961) See p. 863.
5. See Samuel P. Bessman and Judith P. Swazey, "Phenylketonuria: A Study of Biomedical Legislation," Everett Mendelsohn, et al., eds. *Human Aspects of Biomedical Innovation* (Cambridge, MA: Harvard University Press, 1971), p. 53; and Acuff and Faden, "A History of Prenatal and Newborn Screening Programs," p. 64.
6. Acuff and Faden, "A History of Prenatal and Newborn Screening Programs," p. 64.
7. Committee for the Study of Inborn Errors of Metabolism, National Research Council, *Genetic Screening: Programs, Principles, and Research* (Washington, D.C.: National Academy of Sciences, 1975), p. 23.
8. Committee, *Genetic Screening*, p. 50.
9. Judith P. Swazey, "Phenylketonuria: A Case Study in Biomedical Legislation," *Journal of Urban Law 48* (1971), pp. 883–931. See also Diane B. Paul and Paul J. Edelson, "The Struggle over Metabolic Screening," in Soraya de Chadarevian and Harmke Kamminga, eds. *Molecularising Biology and Medicine: New Practices and*

Alliances, 1930s–1970s (Reading: Harwood Academic Publishers, 1988), pp. 203–220.

10. See Bessman and Swazey, "Phenylketonuria," pp. 54–55. See also Katherine L. Acuff and Ruth R. Faden, "A History of Prenatal and Newborn Screening Programs. pp. 59–93, especially pp. 64–65; and Committee, *Genetic Screening*, pp. 44–87.

11. Lori B. Andrews, *State Laws and Regulations Governing Newborn Screening* (Chicago: American Bar Association, 1985), pp. 1–2.

12. Paul J. Edelson, "Lessons from the History of Genetic Screening in the U.S.: Policy Past, Present, and Future," upublished paper, 1995.

13. Paul J. Edelson, "History of Genetic Screening in the United States I: The Public Debate over Phenylketonuria (PKU) Testing," abstract of paper for the American Association for the History of Medicine meeting, New York City, April 28–May 1, 1994.

14. Katherine Bain and Clara Schiffer, *Experience with the Use of PL-480 Funds in Developing PKU Programs in Foreign Countries* (Children's Bureau, Department of Health, Education, and Welfare, 1966).

15. Bain and Schiffer, *Experience with the Use of PL-40 Funds*, p. 7.

16. Bain and Schiffer, *Experience with the Use of PL-40 Funds*, p. 13.

17. Daniel J. Kevles, *In the Name of Eugenics* (New York: Knopf, 1985), p. 254.

18. Marvin R. Natowicz and Joseph S. Alper, "Genetic Screening: Triumphs, Problems, and Controversies," *Journal of Public Health Policy, 12* (1991), p. 479.

19. Collen G. Azen, et al., "Intellectual Development in 12-Year-Old Children Treated for Phenylketonuria," *AJDC*, January 1991, p. 35.

20. Charles R. Scriver, "Phenylketonuria – Genotypes and Phenotypes," (Editorial,) *The New England Journal of Medicine,* May 2, 1991, p. 1280.

21. Jerry E. Bishop and Michael Waldholz, *Genome* (New York: Simon and Schuster, 1990), pp. 18–19.

22. Committee, *Genetic Screening*, pp. 46, 50.

23. Bessman, "Legislation and Advances," p. 337.

24. Committee, *Genetic Screening*, especially pp. 32–40.

25. Committee, *Genetic Screening*, pp. 28–29.

26. Committee, Genetic Screening, p. 34.

27. Neil A. Holtzman, *Proceed with Caution: Predicting Genetic Risks in the Recombinant DNA Era* (Baltimore: The Johns Hopkins University Press, 1989), p. 5.

28. C.C. Mabry, "Phenylketonuria: Contemporary Screening and Diagnosis," *Annals of Clinical and Laboratory Science* November–December. 1990, pp. 393–397.

29. Mark Lappe, *Genetic Politics* (New York: Simon and Schuster, 1979), pp. 92–93.

30. Committee, *Genetic Screening*, pp. 88–89.

31. Harvey L. Levy, "Nutritional Therapy in Inborn Errors of Metabolism," in Robert J. Desnick, *Treatment of Genetic Diseases* (New York: Churchill Livingstone, 1991), p. 16.

32. Azen, "Intellectual Development," pp. 38–39.

33. J. Weglage, et al., "Psychological and Social Findings in Adolescents with Phenylketonuria," *European Journal of Pediatrics* July 1992, pp. 522–525.

34. K. Fishler, et al., "School Achievement in Treated PKU Children," *Journal of Mental Deficiency Research*, December 1989, pp. 493–498.

35. "Phenylketonuria Grows Up" Editorial, p. 1256.

36. See B. Cabalska, et al., "Termination of Dietary Treatment in Phenylketonuria," *European Journal of Pediatrics, 126* (1977) pp. 253–262; and I. Smith, et al., "Effect

of Stopping Low-Phenylalanine Diet on Intellectual Progress of Children with Phenylketonuria," *British Medical Journal, 2* (1978) pp. 723–726. See also M.G. Beasley, "Effect on Intelligence of Relaxing the Low Phenylalanine Diet in Phenylketonuria," *Archives of Disease in Childhood, 66* (March 1991), pp. 311–316.

37. Virginia E. Schuett, et al., "Diet Discontinuation Policies and Practices of PKU Clinics in the United States," *Am J Public Health, 70* (1980) p. 498.

38. Virginia E. Schuett, *National Survey of Treatment Programs for PKU and Selected other Inherited Metabolic Disorders* (Rockville, MD: Bureau of Maternal and Child Health and Resources Development. Public Health Services. U. S. Dept. of Health and Human Services, 1990), p. 11.

39. Susan Waisbren, et al., "The New England Maternal PKU Project: Identification of At-Risk Women," *American Journal of Public Health, 78* (July 1988) pp. 789–792.

40. Waisbren, "The New England Maternal PKU Project," p. 789.

41. Azen, "Intellectual Development," p. 39.

42. Azen, "Intellectual Development," p. 39, citing N.H. Kirkman, "Projections of a Rebound in Frequency of Mental Retardation from Phenylketonuria," *Applied Research in Mental Retardation, 3* (1982) 319–328.

43. "Price List and Order Form," Dietary Specialities, Inc.

44. Betsy A. Lehman, "State Will Aid Those with Enzyme Deficiency," *The Boston Globe*, January 12, 1994, p. 16.

45. "Price List and Order Form."

46. Ellen Wright Clayton, "Screening and Treatment of Newborns," *Houston Law Review* (Spring 1992) p. 94.

47. Committee, *Genetic Screening*, p. 54.

48. Committee, *Genetic Screening*, p. 92.

49. Bernard N. Millner, "Insurance Coverage of Special Foods Needed in the Treatment of Phenylketonuria," *Public Health Reports*, January–February 1993, pp. 60–65.

50. Millner, "Insurance Coverage," p. 64.

51. Clayton, "Screening and Treatment of Newborns," p. 101.

52. See Lappe, *Genetic Politics*, p. 92.

53. Joseph S. Alper and Marvin R. Natowicz, "On Establishing the Genetic Basis of Mental Disease," *Trends in Neurosciences, 16* (1993), pp. 387–389.

54. See for example, essays by Carl Bereiter, "Genetics and Educability: Educational Implications of the Jensen Debate," and by N.J. Block and Gerald Dworkin, "IQ, Heritability, and Inequality" in N.J. Block and Gerald Dworkin, *The IQ Controversy* (New York: Pantheon, 1976), pp. 395–396 and 489; see also Steven Rose, "Environmental Effects on Brain and Behaviour," in Ken Richardson, et al., *Race, Culture and Intelligence* (Baltimore: Penguin Books, 1972), p. 135.

55. Philip Kitcher, *Vaulting Ambition: Sociobiology and the Quest for Human Nature* (Cambridge, MA: MIT Press, 1985), p. 128.

56. Dorothy Nelkin and Laurence Tancredi, *Dangerous Diagnostics: The Social Power of Biological Information* (New York: Basic Books, 1989), p. 160.

57. Nelkin and Tancredi, *Dangerous Diagnostics*, p. 161.

58. Lori B. Andrews, *State Laws and Regulations Governing Newborn Screening* (Chicago: American Bar Association, 1985) p. 2.

59. Nelson and Tancredi, *Dangerous Diagnostics*, pp. 41–42.

60. Ruth Hubbard and Elijah Wald, *Exploding the Gene Myth* (Boston: Beacon Press, 1993), pp. 198–99.

61. Robert Wright, "Dumb bell," *The New Republic*, January 2, 1995, p. 6.

FROM BUTTERFLIES TO BLOOD

Human Genetics in the United Kingdom

DORIS T. ZALLEN

Center for Science and Technology Studies, Virginia Tech, Blacksburg, VA

Without doubt, life was injected into the otherwise backwater field of human genetics in 1953. In that year, the elucidation of the structure of the genetic material, DNA, was accomplished through the collaborative work of James Watson and Francis Crick at the Cavendish Laboratory in Cambridge, England. As it did in other areas of genetics, this scientific insight was to stimulate research in human genetics profoundly, especially through an emphasis on the molecular analysis of DNA and its partners in protein synthesis. At Cambridge, this line of work coalesced into the first research center dedicated to the study of "molecular biology" and became the springboard, four decades later, for human genome projects directed toward identifying and sequencing all of the human genes.

However, this was not the only research effort emanating from the United Kingdom that was to have significant influence on the fields of human and medical genetics. In the mid-1950s, another research partnership was also begun. This collaboration, based on intellectual commitments fostered first at Oxford and later at Liverpool, had a strong organismic and evolutionary focus. It was to yield, among other things, the successful means of dealing with a major medical/genetic problem: Rh-hemolytic disease of the newborn. By engaging the interest and support of the Nuffield Foundation, this work also sparked the establishment of a center for training and research in medical genetics in Liverpool, which produced a generation of the leading human genetics researchers and clinicians in the United Kingdom. It is this second collaboration – between Philip Sheppard, a geneticist principally concerned with inheritance in butterflies and snails, and Cyril Clarke, a physician interested in diagnosing and treating common human diseases – which forms the centerpiece of this article.

Michael Fortun and Everett Mendelsohn (eds.), The Practices of Human Genetics, 197–216
©1999 *Kluwer Academic Publishers. Printed in Great Britain.*

E. B. Ford and the Oxford School of Ecological Genetics

The program of genetic research that developed at Oxford University and was later to influence human genetics bore the unique stamp of one of this century's most unusual scientists: Edmund Brisco ("Henry") Ford. Born in Cumbria in 1901, Ford spent much of his youth engaged in field explorations, studying butterflies with his father and collecting ancient coins and pottery from an old Roman fort that had been situated on the family property. The childhood experience gave him a lifelong love for field work, and he was to make this the foundation of his research methodology when science, instead of the classics, became the dominant interest in his life.

Strongly influenced from his undergraduate days at Oxford and afterward by R. A. Fisher,[1] in his subsequent long and productive research career at Oxford, Ford was to provide the ecological and experimental support for many of Fisher's theoretical ideas about the key role of natural selection in effecting evolutionary change. Ford can be credited with developing a new specialty within genetics that sought to understand detailed mechanisms of evolutionary change by long-term study of features of natural populations (often employing a technique of mark, release, and recapture to quantify the degree of selection taking place in the wild), combined with experimental breeding carried out in the laboratory on some of the specimens he collected. He coined the term "ecological genetics"[2] to describe this type of study and was named professor of ecological genetics in 1963 in recognition of his continued contributions to this area.

Based on the field and laboratory observations he conducted in the 1920s and 1930s, Ford enunciated several concepts which informed all of his subsequent work and which he elaborated on for nearly a half-century in his many books, papers, and private writings. A core element of his research program is his concept of "polymorphism." To be sure, the term polymorphism had been used previously by Darwin to describe natural populations, and had been taken up in genetics by Fisher and others, but in 1940, Ford invested it with a specific meaning that led him to particular predictions about evolutionary processes. Here is his definition:

Polymorphism is...the occurrence together in the same habitat of two or more forms of a species in such proportions that the rarest of them cannot be maintained by recurrent mutation.[3]

The term polymorphism was usually reserved for those situations in which the rarest of the forms observed represented at least 1% of the total population – a level far too high to result from mutation alone. It excluded both those variations that occurred over the course of the life cycle of a single individual and those continuously varying features (such as height) that did not yield clear categories. Thus the occurrence of distinct, heritable color markings in populations of butterflies and snails constituted polymorphisms. In contrast, changes in coloration during the life span of an individual organism or very rare inherited defects whose presence could be accounted for solely by mutation pressure were not polymorphisms in Ford's sense.

Ford went on to argue that, of all the ways that polymorphic variation, which arises from mutation and recombination, can spread and be perpetuated in the population – ultimately reaching a steady-state balance from one generation to the next – natural selection favoring heterozygotes was the most important:

In the first place, it will be realized that any gene which begins to be favoured must exist almost entirely in the heterozygous state during the earlier stages of its increase. That genotype must have gained some superiority over what was the 'normal' homozygote or it would not be displacing it...

Secondly, ...major genes, at least, always have multiple effects and...their mutation occurs at random relative to the needs of the organism.[4]

Ford cautioned against being fooled by how superficial or trivial a particular polymorphism may seem. Each polymorphism, he asserted, has the capacity to yield a variety of effects – not usually detectable by the observer – throughout the metabolic and developmental pathways of the organism. Since some of these effects will be beneficial while others are likely to be harmful,

...the homozygotes will, relative to the alleles in question, have both advantages and disadvantages while the heterozygote will have advantages only, so insuring its superiority over the other two genotypes and therefore establishing polymorphism.[5]

Selective pressures were substantial: by his calculations, some thirty- to fortyfold more powerful than the 1% selective advantage that Fisher himself had estimated as the maximum occurring under natural conditions.

Ford also dismissed any notion that polymorphisms, as he defined them, are neutral, by invoking R.A. Fisher's calculations according to which "the balance of advantage and disadvantage ... must be extraordinarily exact if the two are to behave as neutral in respect of one another."[6] The low probability that such an exquisite balance could be obtained led Ford to reject neutrality as an acceptable explanation.

The underlying genetic state that accounts for these polymorphisms, according to Ford, were "switch-mechanisms."[7] Originally, Ford saw these as arising from the existence of different alleles at a single, major genetic locus, the action of each allele serving as a switch triggering the production of a particular feature or set of features. Later, however, he extended this view by building on the idea of the "super-gene," first expressed in 1949 by Darlington and Mather.[8] Darlington and Mather employed the term super-gene to refer to "a group of genes acting as a mechanical unit in particular allelomorphic combinations" – in other words, a group of genes passed along from one generation to the next in such a way that the individual genes are not separated from one another by the normal processes of recombination. Such a super-gene unit would arise when recombination between genes is inhibited either in a very limited or extended chromosomal region. This restriction on recombination can be caused by the presence of a chromosomal inversion or some other type of mechanical barrier that interferes with normal crossing over or can happen when there are difficulties in achieving normal pairing of homologous chromosomes, as in the case of polyploidy.

Ford found this concept extremely compelling. A cluster of genes inherited as a unit could explain the complex features of many polymorphisms, such as the simultaneous variation in many different flower parts in *Primula*[9] or the color and banding patterns in the snail *Cepaea nemoralis*, as well as the ability to transmit faithfully these sets of characteristics through crosses. And Ford was already quite familiar with an example of such a situation: In 1944, Fisher had offered an interpretation that the Rh blood group system in humans was not a single genetic locus but, in fact, a cluster of three separate but tightly linked genes – C, D, and E – and their respective alleles.[10] In fact, Ford came to regard the production of super-genes, either by the joining together in one location of genes that had been widely dispersed through the genome or through duplication of individual genetic loci, as an ongoing process and an important end product of natural selection. He also considered the super-gene unit as a means of addressing the paradox posed by Mendel's law of independent assortment which, if unchecked, would tend to undermine the orderly development of complicated structures by continuously reassorting the coadapted genes that cooperate in their production. Super-genes would provide a necessary stability and act as a counterweight to the relentless variability imposed by strictly Mendelian mechanisms. Further, the super-genes could not function in isolation but had to be bound into and affected by other genes in the total gene complex. Natural selection acted to favor beneficial changes wrought by modifier genes acting in conjunction with the super-gene switch.

Ford's own program of ecological genetics was mainly concerned with the Lepidoptera (butterflies and moths), but he did have more than a passing interest in human genetics. In particular, he pointed out that the human blood groups were prime examples of balanced genetic polymorphisms.[11] The human blood groups are made up of the various separate genetic systems – such as the ABO blood group locus – which produce unique antigens that reside on the surface of the blood cells. Some of these antigens have to be taken into account when performing blood transfusions; others seem to have little or no significance – medical or otherwise. But they do exist in stable proportions in different population groups throughout the world. Following the line of argument he had developed regarding polymorphisms in general, Ford asserted that these blood group polymorphisms could hardly be neutral but could only have reached their current levels because of a balance of advantages and disadvantages that each confers on the individual. In 1945, he went further and suggested that the observed blood group polymorphisms may be related to susceptibility to specific human diseases:

A valuable line of enquiry which does not yet seem to have been pursued in any detail would be to study blood-group distributions in patients suffering from a variety of diseases. It is possible that in some conditions, infectious or otherwise, they would depart from their normal frequencies, indicating that persons of a particular blood group are unduly susceptible to the disease in question.[12]

However, aside from some brief forays into studies of the taster, color-vision, and secretor polymorphisms, he did not actively pursue this hypothesis himself.

As molecular studies in genetics began to take hold in the 1950s and 1960s in Cambridge and elsewhere, Ford remained staunchly opposed. For him, this new "polymorphism" within biological research had limited vision and no enduring value. And he criticized it in his later work,[13] often cautioning his colleagues against "extolling the virtues of molecular biology,"[14] complaining that "too much genetic research is being put into it,"[15] and labeling the investigators who carried out such work as "incomprehensible interlopers."[16]

There is not enough space here to devote greater detail to the totality of Ford's work or to the many disputes and controversies that it provoked.

Suffice it to say, he created a laboratory at Oxford that exposed numerous undergraduates and a devoted group of postgraduate research workers to the ideas and methodologies of ecological genetics. Prior to 1951, Ford had worked largely on his own. In that year, research benefactors in the form of the Nuffield Foundation[17] and, to a lesser extent, the Agricultural Research Council provided the funding necessary to establish a Genetics Laboratory in the Department of Zoology under Ford's direction. In 1964, the University itself took responsibility for the support of the laboratory,[18] although Ford continued to receive modest Nuffield grants that allowed him to continue his work well into retirement. The laboratory's unique perspective and Ford's fierce advocacy of his views gave it a distinctive stamp and it became known to many simply as the "Oxford School." Individuals such as H.B.D. Kettlewell (who worked on industrial melanism in moths) and Arthur Cain (who studied shell color patterns in the snail *C. nemoralis*) received their research training with Ford, as did Bryan Clarke, Robert Creed, Laurence Cook, David Lees, Kennedy McWhirter, Mark Williamson, and others. But the student who was, through a series of unusual circumstances, to make his mark in human genetics was Philip Sheppard.

The Clarke/Sheppard Collaboration

Born in 1921, Philip Sheppard was not to complete his undergraduate studies at Oxford until 1948, 27 years later, owing to the disruption of World War II, during which he was first in active Royal Air Force service and, for three years, incarcerated in a German prisoner of war camp.[19] His interest in butterflies and flowers soon brought him into the Ford laboratory, and he stayed at Oxford, earning his D.Phil. in 1951 and continuing on there afterward as a junior research officer. It is likely that he would not have had any reason to pursue investigations on natural selection beyond those involving the *Lepidoptera* (especially in *Panaxia dominula*, a moth) and the snail *C. nemoralis*, had it not been for an advertisement he placed in the *Amateur Entomological Society* "Wants and Exchanges" list of October 1952. In that issue, he put in a request for research material:

...living eggs, larvae or pupae of *Papilio machaon* (Swallowtail) from the Continent for genetic research. Will buy, or exchange for living British *P. machaon* or South African *P. demodocus*.[20]

That small advertisement caught the attention of Cyril Clarke, a physician from Liverpool.

Clarke, an internist, was actively involved in clinical practice in internal medicine, but he had an interest in butterflies dating back to his childhood that was rekindled after his second world war naval service in Australia. Intrigued by the mention of the possibility of mating butterflies in E.B. Ford's book, *Butterflies*,[21] Clarke had perfected a method of hand-mating of individual *P. machaon* (swallowtail) butterflies,[22] and he happened to have some spare stock of the hybrid between *P. machaon* and the black U.S. swallowtail *Papilio asterias*[23] that he had recently produced by hand-pairing. He sent some of these hybrids to Sheppard, who found them very interesting, and this led to a series of letters through the following months (October and November 1952), in which they began to exchange ideas, interpretations, food plants, and samples of organisms. By the time they finally met, in December at Oxford, collaborative projects were becoming well thought out. At the start, Sheppard supplied the genetic and statistical expertise and so took the lead in providing the conceptual framework for experiments and in interpreting the results. Clarke, who at first felt that his knowledge of genetics was, in his words, "a little rusty and not up to [Sheppard's] standard,"[24] carried out the experimental manipulations in a greenhouse facility at his own home.

Their joint work grew and expanded to include other areas of investigation in lepidopteran evolution and genetics – especially the genetics of butterfly mimicry – but, for the first few years, it was a collaboration conducted mainly by correspondence.

Clarke's practice was based in Liverpool, where he had obtained a consultant post at the David Lewis Northern Hospital in Liverpool in 1946 and became lecturer in clinical medicine at the University of Liverpool in 1952; Sheppard remained either at Oxford, or at Columbia University, where he spent from September 1954 until July 1955 on a Rockefeller Foundation fellowship with Theodosius Dobzhansky. In 1956, the two partners were finally able to work in closer communication. A position for a senior lecturer in genetics opened up in the Department of Zoology at the University of Liverpool, and Sheppard, whose funding at Oxford was not secure, left Ford's laboratory and moved to Liverpool.[25] Sheppard was to stay there, rising to become the first professor of genetics – a department he formed – until his untimely death from leukemia in 1976 at age 55. Clarke became a reader in the University of Liverpool's Department of Medicine in 1958 and then professor of medicine in 1965, a position he held until his retirement in 1972.

Clarke has often credited Sheppard with teaching him genetics and, based on their earliest years of collaboration, this certainly seems to be a fair assessment. However, expertise also flowed in the reverse direction: Clarke taught Sheppard medicine. The event that appears to have been the catalyst for the expansion of their joint work beyond the *Lepidoptera* and into the area of human and medical genetics was a conversation between Clarke and Sheppard as they were driving out to the Norfolk Broads in East Anglia to collect swallowtail specimens. As the incident has been reported,[26] during the trip Clarke asked Sheppard how his now considerable fund of genetic knowledge could be made use of medically. Sheppard replied immediately: "Blood groups." And from that point on, human blood groups became a major part of their joint research program.

The exact date of this interchange has not been recorded. Clarke recalls that the event occurred sometime during 1953 or 1954. Based on letters between them arranging a trip to the Broads, on the timing of the first appearance of papers on the blood groups in their publication records, and on Sheppard's year-long absence from Britain while at Columbia, that trip probably took place in an interval from the last week of May through early summer of 1953.[27] At the very time that human genetics was being nudged in a new, more molecular direction by the appearance of the Watson and Crick paper on DNA structure,[28] it appears that it was also to be pulled in another by the decision of Clarke and Sheppard to work on the human blood groups.

In retrospect, it is not surprising that Sheppard should have responded to Clarke's query in the way he did or that Clarke endorsed the suggestion enthusiastically. As a student trained in the Oxford School, Sheppard was well acquainted with Ford's view that the human blood groups represented polymorphisms maintained by a balance of selective forces with disease susceptibility being one such possible factor. As a physician, Clarke was certainly aware of the upsurge in blood group research in England at that time following decades of lack of interest.[29] After that discussion on the Norfolk Broads, the decision was made to proceed on two genetic fronts: studies of butterflies and of blood.

Clarke applied the term "family linkage studies"[30] to his first forays into human genetic research, begun while Sheppard was spending the year with Dobzhansky. The term "linkage studies" generally refers to those investigations that seek to build up a map showing the location of genes by determining which genes are positioned so close together on the chromosome that they tend to stay together and are coinherited as they pass from parents to children in families. Actually, what Clarke and his small team at Liverpool had set out to do could more properly be described as *association* studies; that is, they sought to determine if a relationship existed between having particular genes and developing common health problems (for example, conditions of the digestive tract), much as had been reported by Aird *et al.*[31] The several blood group genes then known, including the ABO and Rh genes, were the primary markers used, but there were others as well, including the alleles at the "taster" locus, which were related to the ability to taste a bitter substance called PTC, and the secretor gene locus, whose alleles controlled the ability to produce soluble forms of the ABO blood group antigens in body fluids like saliva and tears. Very quickly, some interesting relationships began to appear that seemed to support Ford's prediction. For example, duodenal ulcer was more likely to occur in blood group O individuals than in individuals of other blood types, although the ulcer association diminished sharply when sib-pairs were substituted for general population studies, as predicted by Lionel Penrose. Further studies were initiated looking for associations of other diseases with specific ABO types, with Sheppard being the one "who mainly designed the work and...did all the statistics."[32] The heterogeneous nature of the disorders chosen and the paucity of markers available made such studies difficult.

The work proceeded in this vein – looking for disease associations with ABO blood types and other genetic markers – until 1958, when it began to move in another, more significant direction. Again, the trigger for this change came from outside of human and medical genetics. It was sparked by findings from Clarke and Sheppard's butterfly research, specifically their work on the genetics of mimicry in the swallowtail butterflies, *Papilio dardanus* and *Papilio memnon*. These are species of butterflies that have evolved to resemble unrelated and highly unpalatable butterflies found in the same region, the mimicry serving to help protect the butterfly from predation by birds. The close physical similarity between one mimetic form of *P. memnon* and the model, *Parides coon*, involves several different features: body color, pattern of pigment distribution on the wings, and the presence of "tails," or protrusions from the hind wings. Doing genetic studies (including, of course, the use of hand-pairing), Clarke and Sheppard found that the genetic basis for this mimicry in *P. memnon* resided in a series of three linked genes, each responsible for a different aspect of the morphologic costume.[33] It was, in fact, a super-gene. Each component gene was responsible for a different aspect of the mimicry. The super-gene was able to be revealed in some gene complexes, but not in others. Thus the presence of the super-gene in females brought about the mimetic changes. The very same super-gene when in males, however, was suppressed. It had no discernible effect, since males never assumed the protective guise.

There was a tantalizing similarity between this type of butterfly genetic organization and one found in humans: the Rh locus. The Rh blood group locus is responsible for the production of a red-blood-cell surface antigen, the Rh factor. The Rh blood group was discovered in 1939 through the separate studies of Levine and Stetson and of Landsteiner and Wiener.[34] The complex serological data that accumulated, when these two groups studied the Rh factor with a variety of different test antisera, led to two different interpretations of gene structure that coexisted for a long period of time in the genetics literature. Wiener accounted for all the confusing data by assuming there was but one Rh locus with a large number of possible alleles; Fisher maintained that the data could be best explained by three linked loci, D, C, and E, each with its own alleles.[35] As noted previously, the Rh locus was one of the first examples of a possible "super-gene" on record.

Sheppard, as a disciple of the Oxford School, was well aware of and accepted the super-gene interpretation.[36] Clarke, as a physician, was well aware of the medical importance of the Rh blood groups. The blood-type incompatibility that exists between an Rh-negative woman carrying an Rh-positive fetus can lead to the production of strong immune system reactions by the mother against the Rh antigen. Once the Rh-negative mother has been sensitized, subsequent Rh-positive fetuses will be in jeopardy, causing them to suffer from the damaging effects of the destruction of their red blood cells – an often fatal condition known as erythroblastosis fetalis or Rh disease. The strong similarity between the Fisher interpretation of the structure of the Rh locus and the organization of the butterfly super-gene stimulated Clarke and Sheppard to decide to go beyond ABO studies and make the Rhesus blood group and Rh disease a major part of their program. In fact, their research and that of their coworkers began to move quickly in this new direction.

Within a few years, that line of investigation led to the means of preventing Rh disease. The entire story, which cannot be presented here, is regarded as a major medical and genetic research success, involving groups in the United States as well as in the United Kingdom.[37] The specific approach that the Liverpool workers took was informed by their ecological genetic framework, one tenet of which is that the action of major genes or super-genes are connected to and affected by the action of other genes residing elsewhere in the total gene complex. For the Liverpool researchers, the key was the observation by Levine[38] that ABO incompatibility between mother and fetus offered substantial protection from Rh incompatibility. This interaction between the ABO locus and the

Rh locus was seen as probably the result of the expeditious removal of fetal red blood cells from the maternal circulation by naturally occurring maternal antisera to A and B antigens (in type O women) before any immune system response could be mounted against the Rh antigen on those fetal cells.

The Liverpool program sought to verify Levine's report, which it did in 1958.[39] Then it sought some ways of medically mimicking the effect of the ABO locus in women at risk. This the Liverpool researchers accomplished by a feat of fancy pharmacologic footwork that first originated in an idea put forth by Ronald Finn, a physician and researcher in Clarke's group, at a genetics meeting that Clarke had organized at the Liverpool Medical Institution.[40] The plan was to foil the mother's immune system by injecting her with a small dose of antisera to the most potent Rh-complex gene, D, shortly after the delivery of an Rh-positive child. If the antiserum rapidly and completely destroyed any Rh-positive cells that might have crossed the placenta and entered the mother's blood, her own immune system would not be able to detect any fetal Rh antigen and therefore would not be stimulated to produce antibodies against the Rh factor, which might harm fetuses in future pregnancies. To this day, this same technique has virtually eliminated Rh disease.

In the early 1960s, the Liverpool group was perfecting the details of their treatment scheme and starting their human trials. Their work attracted attention that was not only to enable them to proceed more efficiently on the Rh problem but was also to enlarge the entire medical genetic research effort at Liverpool and allow it to become a focal point for research and training in medical genetics in the United Kingdom.

The Nuffield Foundation: Support and Stability for Medical Genetics

From its start in 1943, the Nuffield Foundation had provided funding for a diverse set of research programs in the biological sciences and medicine. These programs included the establishment and support of E.B. Ford's laboratory, as well as of Philip Sheppard's butterfly work while at Oxford and after his move to Liverpool.[41] The justification for this support probably resided in the perceived relevance of the genetic studies to medicine, although precise medical applications were never spelled out.[42] Since butterfly research could be conducted quite inexpensively, the Nuffield grants were similarly modest in scope. This was to change in 1963, when the Nuffield Foundation gave one of its largest grants ever –

£350,000 – to establish a Unit of Medical Genetics within the Department of Medicine at the University of Liverpool.[43]

The pivotal person in bringing this about was the late Dame Janet Vaughan. Vaughan, a physician specializing in the treatment of blood disorders and with expertise in blood transfusion, became principal of Somerville College at Oxford in 1945. She was also one of the original trustees of the fledgling Nuffield Foundation.[44] She held both posts for more than 20 years. Her medical and scientific expertise, along with her knowledge of the academic community, gave her a position of some authority within the Foundation and enabled her to take the lead in guiding many of its funding decisions. According to John McAnuff, assistant director of the Nuffield Foundation, it was Dame Janet's inspiration to direct substantial resources to the establishment of a center for medical genetics, and it was she who vigorously lobbied the other trustees on its behalf.[45]

According to Ford's recollections of the unit's origins,[46] the first approach was made to him:

The Foundation was uncertain where, if at all, to place such a unit; Oxford was briefly considered and they asked me to be the Director of it. I said that was quite impossible: The Director *must* be a medical man and at that time there was only one person in the country for the job – [Clarke], so that, taking all into consideration, such a unit *must* be cited (sic) at Liverpool....When I put these points to Janet V. she accepted them both without question and did much to get the Foundation to do so too.

Janet Vaughan was a close colleague of E. B. Ford at Oxford and was also aware of his links to the blood group genetics research being carried out by Clarke and Sheppard. As a long-standing grantee and a leading geneticist, Ford commanded a great deal of respect within the foundation and was often called up to act as a referee in its grants program. In this case, not only was his advice heeded but he was also pressed into service as an emissary of the foundation to explore the possibility of support for medical genetics at Liverpool.

The negotiations started at the beginning of 1963 and were to go on for nearly the whole of that year. From the start, Vaughan kept Foundation Director Leslie Farrer-Brown apprised of the progress:

I had a long session with Ford yesterday [February 18, 1963] and we planned that he should work out some scheme with Shepherd [sic] and Clark [sic] at Liverpool, and put it up to the Foundation; the plan being that there should be a close association in anything done between the Oxford and Liverpool groups, but that the medical side of it would, of course, be done in Liverpool. I think this holds out good promise for the future, building on what the Foundation has already done and spreading ourselves into one of the most promising fields in Medicine. I think and hope that I encouraged Ford to think in a big way in the first instance – it is always easier to prune than to expand.

Vaughan then added a note of urgency:

...I have seen the new plans for the Liverpool Teaching Hospital and of course there is no mention of Medical Genetics therein. The plans are not complete, however, and it would I think be helpful if we could insinuate something of what we want before it is too late. I have a sort of feeling that something really very exciting is going to grow out of all this.[47]

Over the next several months, Ford met first with Sheppard and then with Clarke. Officers of the foundation and Janet Vaughan then continued the dialogue. The Clarke and Sheppard response to the opportunity was enthusiastic.

Some requirements emerged early on: The unit had to have some real affiliation with the university in order to allow it to introduce medical genetics into the training of the physicians. There also had to be a way to assure direct contact with patients. Ford reported that:

...access to patients...would be completely fundamental to the success of such an Institute. It also provides the right "entry" into the Liverpool hospitals. The fact that neither the Galton Laboratory in London nor the M.R.C. [Medical Research Council] place at Headington have hospital beds attached to them is one of the reasons why they have both fallen completely behind what is already being done at Liverpool....[48]

The plan that evolved over this period, and which was ultimately approved on October 4, 1963 by the trustees, was for one of the largest grants ever awarded by the Nuffield Foundation – £350,000 – over nine years, to establish the Nuffield Unit of Medical Genetics within the Liverpool University Department of Medicine. A large portion was to be used for the construction of a building to house the many different facets of the research program. The rest was to hire additional researchers, fund fellowships that would bring clinicians to the unit to study genetic aspects of human disease, and to pay some of the operating costs of the new facility. The unit, with Clarke at the helm as director, was to work closely with the basic research efforts in ecological genetics being carried out by Sheppard in the Department of Genetics and with Ford's group at Oxford. It was to have extensive clinical contact with the patients in the associated teaching hospitals and the ability to reach out to the population at large.

The unifying thread of the entire research program was the polymorphism concept. Each of the proposed component programs was designed to identify human polymorphisms, investigate their relationship to human disease, and develop treatments that could outwit the negative selective forces involved.[49] Several areas were to be emphasized: the prevention of Rh disease; the search for additional polymorphisms related to blood factors or the ability to metabolize drugs; the cytological study of human chromosomes for structural evidence of inversion polymorphisms; and the study of immunological polymorphisms predisposing to cancer. The concern throughout was for studying the genetic basis of the major types of human disorders: gastrointestinal diseases, cancer, thyroid disease, schizophrenia, blood defects, Rh disease, rheumatoid arthritis (this was probably included because it was of special concern to Lord Nuffield), and the like.

Once the grant was awarded, medical genetics research at Liverpool proceeded at an accelerated pace within the newly defined unit. Sheppard continued his insect genetics work within the Department of Genetics while playing a key role in assisting in the design and interpretation of many of the experiments conducted in the Nuffield Unit. On weekends, he and Clarke met at Clarke's home to work on their butterfly genetics projects. Within two years, support for Sheppard's work was brought under the auspices of the large Nuffield grant.[50] Although Sheppard and Clarke were in touch with Ford about various aspects of the butterfly experiments, Ford played only a peripheral part, at best, in the human genetics work. Perhaps his suggestion, some years earlier, that people with AB blood type be completely transfused with type O blood revealed such a lethal lack of medical sophistication that he could not be trusted with a greater role in clinical matters.[51]

The building housing the new unit was completed in early 1967 and was dedicated on May 26, 1967. Ford was given the honor of officially opening the building in gratitude for his intervention with the Nuffield Foundation. The building had all the usual features one would associate with a medical research facility of that period; the only highly unusual note was the presence of a greenhouse on the roof to house butterfly experiments.

The Influence of the Oxford School and the Nuffield Unit on Medical Genetics

The generous support garnered from the Nuffield Foundation allowed the ideas of the Oxford School, as applied in genetic medicine, to reach a much wider and more diverse audience than would ever have been possible from a single research program alone, no matter how successful. From the early 1960s to the 1970s, an increasingly diverse set of research activities, involving a range of researchers from scientific and clinical areas, took place under the auspices of the Nuffield Unit of Medical Genetics. The unit also had an educational component that provided training in genetics to scientists and physicians.

A number of the research efforts conducted there have borne fruit. Most prominent of these was the demonstration that anti-D gamma-globulin administration does, indeed, protect against Rh disease. By 1967, the Liverpool researchers had gone beyond the trials on Rh-negative policemen and postmenopausal women to its use on pregnant women at highest risk.[52] Their work, in conjunction with that of groups in the United States, has brought an end to Rh disease in all newborns of properly treated Rh-negative women. Another Liverpool program, led chiefly by David Price Evans, identified an enzyme polymorphism that allows some individuals to acetylate certain drugs rapidly, while others can carry out such reactions only at very slow rates. The recognition of the genetic basis of differential response to drugs has implications for the selection of proper levels of drugs in medical treatment. It has also spurred the growth of a new genetic specialty: pharmacogenetics. Another successful avenue of investigation involved the pursuit of hemoglobin polymorphisms, which in turn, has led to the understanding of the genetic details of thalassemia and other blood disorders.

The education and training aspect of the Nuffield Unit at Liverpool has also left its mark on clinical genetics in the United Kingdom. Many of those currently heading the leading medical genetics centers are alumni of its program. David Weatherall, who has done groundbreaking work on the genetics of hemoglobin disorders, now is directing the Department of Clinical Medicine in the Faculty of Clinical Medicine at Oxford University. Other intellectual descendents include Marcus Pembrey at the Institute of Child Health in London, Peter Harper at Cardiff, and Rodney Harris at Manchester – and David A. Price Evans, John Woodrow, and several others who are now retired.

Effects on the training of physicians also came from the medical genetics texts of Ford and, later, Clarke.[53] Unlike other texts, these books are unique because they make polymorphisms and the associated medical implications a central organizing principle throughout. By the communication of these genetic ideas to medical students in their courses and the widespread use of these texts, the Oxford School and its Liverpool outpost were able to extend their genetic vision into the community of practicing physicians. With it has come an emphasis on the importance of genetics for understanding *major* health problems, not just rarer types of single-gene disorders that had typically occupied the attention of medical geneticists; the notion that susceptibility to a disorder can be conferred by a particular set of genes and that these genes can be used to determine who might be predisposed to an illness in advance of any symptoms; and the recognition that the whole genetic endowment, not just an isolated locus, must be considered in constructing genetic studies and fashioning treatments.

Many of Ford's claims about polymorphisms and super-genes have not withstood the test of time. With the advent of recombinant DNA technology, the term polymorphism has taken on a powerful new meaning in genetics, leading to the use of restriction fragment length polymorphisms (RFLPs) in genetic testing and acting as a means of probing the DNA molecule. Many of Clarke's and Sheppard's genetic claims have been superseded by later research. And the Nuffield Unit no longer exists, having been taken over in bits and pieces by other medical units. Nonetheless, the legacy of the Oxford School and the Nuffield butterfly workers and clinicians is still being felt in human and medical genetics.

Acknowledgements

This work was greatly aided by Colin Harris, Heather Creamer, and S.R. Tomlinson in the Department of Western Manuscripts of the Bodleian Library, Oxford, who provided access to the E.B. Ford papers; Thomas Rosenbaum, archivist, Rockefeller Foundation archives, Tarrytown, New York, who tracked down Philip Sheppard materials; and the Nuffield Foundation, London, which generously made available their records of the Nuffield Unit of Medical Genetics. Sir Cyril Clarke kindly consented to several interviews, and Alison Gill patiently showed me the Clarke and Sheppard butterfly collection at the Natural History Museum, London. A travel grant from the Wellcome Trust and the hospitality of William Bynum and Tilli Tansey and the staff at the Wellcome Institute for the History of Medicine in London made the research and writing possible.

Notes and References

1. Ford was a student of Julian Huxley at Oxford. After hearing about Ford from Huxley, Fisher travelled to Oxford to meet him. Arriving unexpectedly one day at Wadham College, Fisher was obliged to wait for some time until Ford eventually returned to his rooms and discovered his eminent visitor. This episode is recounted in papers from the Edmund Brisco Ford collection held at the Bodleian Library, Oxford (herein referred to as the Ford papers, Bodleian), Ms Eng., c. 2646, A.9. Despite their different professional affiliations, Fisher and Ford were to work together from 1923 until Fisher's death in 1963.
2. Ford first planned to write a book on ecological genetics, or the study of evolution by the coordinated use of ecological and genetic investigations, in 1928 – choosing, at that time, to wait until he had conducted sufficient research. He had originally thought that would take no more than about 25 years. It took him 36 years to complete the volume that codified the field: E.B. Ford, *Ecological Genetics* (London: Methuen, 1964).
3. The concept of polymorphism is fleshed out in E.B. Ford, "Polymorphism and Taxonomy," in *The New Systematics* ed., J. Huxley (London: Oxford University Press, 1940), pp. 493–513. However, the more succinct definition given here appears in *Genetics for Medical Students* by Ford E.B. (London: Methuen & Co, Ltd., 1942). See also: E.B. Ford, *Genetic Polymorphism* (London: Faber and Faber, 1965), p. 11.
4. Ford, *Genetic Polymorphism*, pp. 26–27.
5. Ford, *Genetic Polymorphism*, p. 27.
6. Ford, *Genetic Polymorphism*, pp. 12–13.
7. Ford, *Genetic Polymorphism*, p. 11.
8. C.D. Darlington and K. Mather, *The Elements of Genetics* (London: Allen and Unwin, 1949).
9. In the primrose, *Primula*, the "thrum" and "pin" phenotypes are the result of a number of different genes affecting the height of the pollen-producing anthers, the size of the pistil, and the rate of germination of pollen grains. The genes for these separate features are so closely linked that they appear to be inherited as a single unit, i.e., a super-gene. Ford, *Ecological Genetics*, pp. 172–185.
10. R.A. Fisher, cited by R.R. Race, "An 'Incomplete' Antibody in Human Serum," *Nature*, *153* (1944), pp. 771–772.
11. Ford, *Genetics for Medical Students*, pp. 103–111.
12. E.B. Ford, "Polymorphism," *Biological Reviews of the Cambridge Philosophical Society*, *20* (1945), p. 85.
13. Ford, *Genetic Polymorphism*, pp. 28–29.
14. Ford papers, Bodleian: Ms Eng. c. 2660, F. 108: letter from Ford to P.M. Sheppard, February 12, 1970.
15. Ford papers, Bodleian: Ms Eng., c. 2660, F.97: letter from Ford to P.M. Sheppard, June 21, 1968.
16. Cyril Clarke, "Professor E.B. Ford" (Obituary), *The Independent*, January 25, 1988.
17. The Nuffield Foundation was the largest and most prominent of the several philanthropic enterprises endowed by William Morris (Lord Nuffield), designer of the Morris motor car and founder of Morris Motors Ltd. The Foundation's stated purposes were threefold: "the advancement of health and the prevention and relief of

sickness [through] medical research and teaching," "the advancement of social well being," and "the comfort and care of the aged poor." See Ronald W. Clark, *A Biography of the Nuffield Foundation* (London: Longman Group Ltd, 1972), pp. 8–9.

18. Ford papers, Bodleian: Ms Eng., c. 2660, F.82: letter from Ford to Mr. Paterson, Secretary of Faculties, University Registry, Oxford, February 25, 1964.
19. Fuller biographical information can be found in Cyril Clarke, "Philip MacDonald Sheppard, 1921–1976," *Biographical Memoirs of Fellows of the Royal Society, 23* (1977), pp. 464–500.
20. Amateur Entomological Society (A.E.S.), Wants and Exchanges List No. 18, October 1952, p. 3.
21. E.B. Ford, *Butterflies* (London: Collins, 1945).
22. C.A. Clarke, "Hand Pairing of *Papilio machaon* in February," *Entomologist's Record, 64* (1952), pp. 98–100.
23. C.A. Clarke and J.P. Knudsen, "A Hybrid Swallowtail (an Account of the Cross *Papilio asterias* x *Papilio machaon* and a Note on the *Machaon* Complex of the North American continent)," *Entomologist's Record, 65* (1953), pp. 76–80. See also C.A. Clarke, letter to P.M. Sheppard, October 2 1952, in Clarke-Sheppard collection at the Museum of Natural History, London (herein referred to as the C-S collection, London).
24. C.A. Clarke, letter to P.M. Sheppard, October 14, 1952, C-S collection, London.
25. Cyril Clarke has pointed out that his knowledge of people, honed by his years of clinical practice, were useful in attracting Sheppard to Liverpool. In order to encourage him to consider the new post, Clarke decided to use a bit of reverse psychology, mentioning some of the drawbacks and asking Sheppard to call this opportunity to the attention of anyone else who would be interested. Within a short time, Sheppard himself applied. See C.A. Clarke, "Philip MacDonald Sheppard, 1921–1976," *Biographical Memoirs of the Royal Society, 23* (1977), pp. 465–500.
26. This incident is mentioned in several places, including Clarke, "Philip MacDonald Sheppard," p. 486 (note r); C.A. Clarke, Royal College of Physicians and Oxford Polytechnic Video Archive, c. 1986; and C.A. Clarke, interviews, May 21, 1991 and May 26, 1993.
27. C.A. Clarke letters to P.M. Sheppard, April 27, 1953 and May 11, 1953 and P.M. Sheppard, letter to C.A. Clarke, May 20, 1953, both C-S collection, London. This correspondence indicates that a trip took place in late May 1953. In a recent communication, Cyril Clarke suggests that the collecting trip may have taken place a little later – in early summer – because swallowtails emerge in June.
28. J.D. Watson and F.C. Crick, "Molecular Structure of Nucleic Acids: A Structure for Deoxyribose Nucleic Acids," *Nature, 171* (1953), pp. 737–738.
29. William H. Schneider, "British Research on the Genetics of Human Blood Groups Between the Wars," paper for the American Association for the History of Medicine meeting Cleveland, Ohio, May 2, 1991.
30. C.A. Clarke, Letter to L.S. Penrose, March 9, 1954; Henry Cohen, letter to L.S. Penrose, April 12, 1954; and C.A. Clarke, letter to L.S. Penrose, May 24, 1954. (All items are in the Penrose papers 123/7, held at University College, London). During this phase of his move into human genetic studies, Clarke turned for advice to the serological genetics group, established by Fisher at the Galton Laboratory in London before the second world war. The Galton professor at that time was Lionel S. Penrose.

31. I. Aird, H.H. Bentall, and J.A. Fraser Roberts, "A Relationship Between Cancer of Stomach and the ABO Blood Groups," *British Medical Journal, 1* (1953), pp. 799–801.
32. Clarke, "Philip MacDonald Sheppard," p. 487.
33. C.A. Clarke and P.M. Sheppard, "The Genetics of *Papilio dardanus* Brown I. Race *Cenea* from South Africa," *Genetics, 441* (1959), pp. 1347–1358; C.A. Clarke and P.M. Sheppard, "Further Studies on the Genetics of the Mimetic Butterfly *Papilio memnon* L.," *Philosophical Transactions, Royal Society London B., 263* (1971), pp. 45–76.
34. P. Levine and R. Stetson, "An Unusual Case of Intra-Group Agglutination," *Journal of the American Medical Association, 113* (1939), pp. 126–127; K. Landsteiner and A.S. Wiener, "An Agglutinable Factor in Human Blood Recognized by Immune Sera for Rhesus Blood," *Proceedings of the Society for Experimental Biology and Medicine, 43* (1940), p. 223.
35. Race, "An 'Incomplete' Antibody in Human Serum," p. 772. A more complete description of Fisher's reasoning can be found in R.A. Fisher, "The Rhesus Factor: A Study in Scientific Method," *American Scientist, 35* (1947), p. 113.
36. P.M. Sheppard, "Polymorphism, Linkage and the Blood Groups," *The American Naturalist, 87* (1953), p. 291; P.M. Sheppard, "The Rh Blood-Groups," *Lancet, i* (1957), p. 212.
37. There have been many recountings of the entire Rh story, from the identification of the basis of the disease in newborns to its treatment and prevention. See, for example, David R. Zimmerman, *Rh: The Intimate History of a Disease and its Conquest* (New York: Macmillan Publishing Co, Inc., 1973).
38. P., Levine, "Serological Factors as Possible Causes in Spontaneous Abortions," *Journal of Heredity, 34* (1943), pp. 71–80.
39. C.A. Clarke, R. Finn, R.B. McConnell, and P.M. Sheppard, "The Protection Afforded by ABO Incompatibility Against Erythroblastosis due to Rhesus Anti-D," *International Archives of Allergy and Applied Immunology, 13* (1958), p. 380.
40. Ronald Finn, "Erythroblastosis", a report of the Symposium on the Role of Inheritance in Common Diseases in Liverpool Medical Institution, February 18, 1960, *Lancet, i* (1960), p. 526.
41. Clark, *A Biography of the Nuffield Foundation*, p. 187.
42. Clark, *A Biography of the Nuffield Foundation*, p. 187.
43. Minutes, 130th Meeting (October 4, 1963) of the trustees in *Minutes of the Meetings of the Trustees (and Annexed Papers), 1963–4* (London: Nuffield Foundation, 1964). In 1969, a supplementary award of £39,200 was made.
44. For biographical information, see Helen Dodsworth, "Dame Janet Vaughan (1899-1993)" in Lynn Bindman, Alison Brading, and Tilli Tansey, eds., *Women Physiologists* (London: Portland Press, 1993), pp. 31–36.
45. John McAnuff, interview, July 25, 1991.
46. Ford papers, Bodleian: Ms Eng., c. 2656, F.21. E.B. Ford, letters to C.A. Clarke, May 11, 1977 and May 14, 1977.
47. Janet Vaughan, letter to L. Farrer-Brown, February 19, 1963, Med/91, Q5/2/1, Nuffield Foundation, London (hereafter referred to as NF).
48. E.B. letter to L. Farrer-Brown, April 8, 1963, Med/91, Q5/2/1 (NF).
49. See E.B. Ford, letter to L. Farrer-Brown, May 27, 1963; and C.A. Clarke, letter to J. McAnuff, September 11, 1963, Med/91, Q5/2/1 (NF).
50. W.H.F. Barnes, letter to B. Young, March 30, 1965, Med/91, Q5/2/1 (NF).

51. Ford, "Polymorphism", p. 85.
52. C.A. Clarke, "Prevention of Rh-Haemolytic Disease," *British Medical Journal*, 4 (1967), pp. 7–12.
53. Ford's *Genetics for Medical Students* went into six editions from the 1940s until the 1960s, C.A. Clarke, *Genetics for the Clinician* (Oxford: Blackwell Scientific Publications, 1962); C.A. Clarke, *Human Genetics and Medicine* (London: Arnold, 1st ed., 1970, 3rd ed., 1987).

Sociology of the Sciences

1. E. Mendelsohn, P. Weingart and R. Whitley (eds.): *The Social Production of Scientific Knowledge.* 1977   ISBN Hb 90-277-0775-8; Pb 90-277-0776-6
2. W. Krohn, E.T. Layton, Jr. and P. Weingart (eds.): *The Dynamics of Science and Technology.* Social Values, Technical Norms and Scientific Criteria in the Development of Knowledge. 1978   ISBN Hb 90-277-0880-0; Pb 90-277-0881-9
3. H. Nowotny and H. Rose (eds.): *Counter-Movements in the Sciences.* The Sociology of the Alternatives to Big Science. 1979
      ISBN Hb 90-277-0971-8; Pb 90-277-0972-6
4. K.D. Knorr, R. Krohn and R. Whitley (eds.): *The Social Process of Scientific Investigation.* 1980 (1981)   ISBN Hb 90-277-1174-7; Pb 90-277-1175-5
5. E. Mendelsohn and Y. Elkana (eds.): *Sciences and Cultures.* Anthropological and Historical Studies of the Sciences. 1981
      ISBN Hb 90-277-1234-4; Pb 90-277-1235-2
6. N. Elias, H. Martins and R. Whitley (eds.): *Scientific Establishments and Hierarchies.* 1982   ISBN Hb 90-277-1322-7; Pb 90-277-1323-5
7. L. Graham, W. Lepenies and P. Weingart (eds.): *Functions and Uses of Disciplinary Histories.* 1983   ISBN Hb 90-277-1520-3; Pb 90-277-1521-1
8. E. Mendelsohn and H. Nowotny (eds.): *Nineteen Eighty Four: Science between Utopia and Dystopia.* 1984   ISBN Hb 90-277-1719-2; Pb 90-277-1721-4
9. T. Shinn and R. Whitley (eds.): *Expository Science.* Forms and Functions of Popularisation. 1985   ISBN Hb 90-277-1831-8; Pb 90-277-1832-6
10. G. Böhme and N. Stehr (eds.): *The Knowledge Society.* The Growing Impact of Scientific Knowledge on Social Relations. 1986
      ISBN Hb 90-277-2305-2; Pb 90-277-2306-0
11. S. Blume, J. Bunders, L. Leydesdorff and R. Whitley (eds.): *The Social Direction of the Public Sciences.* Causes and Consequences of Co-operation between Scientists and Non-scientific Groups. 1987
      ISBN Hb 90-277-2381-8; Pb 90-277-2382-6
12. E. Mendelsohn, M.R. Smith and P. Weingart (eds.): *Science, Technology and the Military.* 2 vols. 1988
      ISBN Vol, 12/1 90-277-2780-5; Vol. 12/2 90-277-2783-X
13. S. Fuller, M. de Mey, T. Shinn and S. Woolgar (eds.): *The Cognitive Turn.* Sociological and Psychological Perspectives on Science. 1989
      ISBN 0-7923-0306-7
14. W. Krohn, G. Küppers and H. Nowotny (eds.): *Selforganization.* Portrait of a Scientific Revolution. 1990   ISBN 0-7923-0830-1
15. P. Wagner, B. Wittrock and R. Whitley (eds.): *Discourses on Society.* The Shaping on the Social Science Disciplines. 1991   ISBN 0-7923-1001-2

Sociology of the Sciences

16. E. Crawford, T. Shinn and S. Sörlin (eds.): *Denationalizing Science. The Contexts of International Scientific Practice.* 1992 (1993)   ISBN 0-7923-1855-2
17. Y. Ezrahi, E. Mendelsohn and H. Segal (eds.): *Technology, Pessimism, and Postmodernism.* 1993 (1994)   ISBN 0-7923-2630-X
18. S. Maasen, E. Mendelsohn and P. Weingart (eds.): *Biology as Society? Society as Biology: Metaphors.* 1994 (1995)   ISBN 0-7923-3174-5
19. T. Shinn, J. Spaapen and V. Krishna (eds.): *Science and Technology in a Developing World.* 1995 (1997)   ISBN 0-7923-4419-7
20. J.Heilbron, L. Magnusson and B.Wittrock (eds.): *The Rise of the Social Sciences and the Formation of Modernity.* Conceptual Change in Context, 1750–1850. 1996 (1998)   ISBN 0-7923-4589-4
21. M. Fortun and E. Mendelsohn (eds.): *The Practices of Human Genetics.* 1998
                ISBN 0-7923-5333-1

KLUWER ACADEMIC PUBLISHERS – DORDRECHT / BOSTON / LONDON